Assertions and Refutations VI

The Poupard "Solution"
of the Galileo Case

or

As a true elect, the Holy Father will never be deceived nor will he ever be a deceiver

An Historical Analysis

Frits Albers, PH.B.
May 1993 / December 1999

Edited and with a Preface
by
Frank Calneggia

Make the time

En Route Books and Media, LLC
5705 Rhodes Avenue
St. Louis, MO 63109

Contact us at contactus@enroutebooksandmedia.com

Cover: A bust of St. Robert Bellarmine, an Italian Jesuit cardinal who was one of the Church's leading scholars during the Counter-Reformation, executed by the Italian artist Gian Lorenzo Bernini in the years 1621–1624, and unveiled in August 1624. It is in the Chiesa del Gesù, Rome. It was commissioned by Pope Gregory XV and Cardinal Odoardo Farnese after Bellarmine's death in 1621.

Copyright 2024 Michael P. Albers
ISBN: 979-8-88870-291-8 and 979-8-88870-296-3
Library of Congress Control Number: 2024951978

The address given by His Holiness Pope John Paul II to the members of the Pontifical Academy of Sciences on October 31, 1992, printed in the *Osservatore Romano*, November 4, 1992, is reprinted with permission. © Dicastero per la Comunicazione-Libreria Editrice Vaticana.

All rights reserved. No part of this book may be reproduced, stored in a retrieval system, or transmitted in any form, or by any means, electronic, mechanical, photocopying, or otherwise, without the prior written permission of the author.

Dedicated to
St. Robert Bellarmine
(4 October 1542 – 17 September 1621)

St. Robert Bellarmine was the third of ten children of a noble but impoverished family. His mother Cinzia Cervini was a niece of Pope Marcellus II. His childhood was marked by ill health. It was feared he would not live to reach manhood.

He entered the newly formed Society of Jesus in 1560 and began a systematic study of theology. In 1569 he was sent to Leuven (Louvain) to study the prevailing heresies of his time. He was ordained there and became famous for his Latin sermons. He was the first Jesuit to teach at the university. His course was the *Summa* of Thomas Aquinas. He taught there for seven years.

In 1576 St. Robert journeyed to Rome where Pope Gregory XIII commissioned him to teach polemical theology at the new Roman College. He taught for eleven years and also wrote his *Disputationes* which was a compilation of the various controversies of his time. This work had an immense affect on Protestants and to this day it is still the classic treatise on the subject.

He was made Rector of the Roman College in 1592, Provincial of Naples in 1594, examiner of Bishops in 1598 and Cardinal in 1599. Pope Clement VIII said of him "the Church of God had not his equal in learning". Immediately after his appointment as Cardinal, Pope Clement made him a Cardinal Inquisitor.

St. Robert Bellarmine is noted for his work during the *Counter-Reformation* in which he implemented the decrees of the Council of Trent. He was famous throughout all of Europe as a theologian and as a strenuous defender of the Faith. His prolific writing included works of instruction and devotion. During his retirement he wrote several short books intended to help ordinary people in their spiritual life.

St. Robert was the confessor, spiritual father and friend of St. Aloysius Gonzaga whose cause for beatification he promoted. He helped St. Francis de Sales obtain formal approval of the Visitation Order.

He lived a very simple and humble life. He lived on the bare essentials and only ate the food available to the poor. St. Robert seemed happiest when he was working with the poorest of the poor to whom he would give his own personal goods and spiritual comfort. He had a special devotion to St. Francis of Assisi on whose Feast Day he had been born. He died on the Feast Day of the Stigmata of St. Francis, 17th September.

His remains, in a cardinal's red robes, are displayed behind glass under a side altar in the Church of Saint Ignatius, the chapel of the Roman College, next to the body of his student Aloysius Gonzaga, as he himself had wished.

Pope Pius XI beatified Roberto Francesco Romulo Bellarmino in 1923, canonised him in 1930 and declared him a Doctor of the Church in 1931.

Quotations

I
St. Thomas Aquinas

Reason may be employed in two ways to establish a point: firstly, for the purpose of furnishing sufficient proof of some principle, as in natural science, where sufficient proof can be brought to show that the movement of the heavens is always of uniform velocity. Reason is employed in another way, not as furnishing a sufficient proof of a principle, but as confirming an already established principle, by showing the congruity of its results, as in astrology [i.e. astronomy] the theory of eccentrics and epicycles is considered as established, because thereby the sensible appearances of the heavenly movements can be explained; not, however, as if this proof were sufficient, forasmuch as some other theory might explain them. (ST. I q.32, ad 2)

II
Johan Ludvig Emil Dreyer
(John Louis Emil Dreyer)

J.L.E. Dreyer. Ph.D. *History of the Planetary Systems From Thales to Kepler.* Cambridge University Press. 1906.

Dreyer's book is still considered to be the standard reference on the subject in the English speaking world. The author was a Danish astronomer who worked for the greater part of his professional life in Ireland. He was director of Armagh Observatory from 1882 until his retirement in 1916, in which year he moved to Oxford and edited the works of the great Danish astronomer Tycho Brahe. He died in 1926.

A number of the quotations I have chosen from Dreyer contain explanations and discussions of geometrial figures. I have not reproduced these geometrial figures. In order to retain the fluidity and ordered construction of these particular texts it was necessary to retain the geometrical and mathematical explanations and discussions. This notwithstanding the main facts and conclusions in these quotations are still clear. As an aid I have added underling to emphasise them.

Quotations of Dreyer

Aristarchus is the last prominent philosopher or astronomer of the Greek world who seriously attempted to find the physically true system of the world. After him we find various ingenious mathematical theories which represented more or less closely the observed movements of the planets, but whose authors by degrees came to look on these combinations of circular motion as a mere means of computing the position of each planet at any moment, without insisting on the actual physical truth of the system. Three names stand out clearly among the astronomers of the next four hundred years as the principal, or perhaps we should say the only, promoters of theoretical astronomy: Apollonius (B.C. 230), Hipparchus (B.C. 130), and Ptolemy (A.D. 140). (p. 149)

In the noble dedication to Pope Paul III with which his book opens, Copernicus says that he was first induced to seek for a new theory of the heavenly bodies by finding that mathematicians differed greatly among themselves on this subject. (p. 311)

By giving an annual orbit to the sun and making it account for the "second inequalities", Copernicus had laid the foundation of a system very much simpler than the Ptolemaic system. But unfortunately he was compelled to mar the simplicity of his work, because the heliocentric system was not sufficient to explain the varying velocities of the planets in their orbits, the "first inequalities." There was no help for it, he had to make use of the excentrics and epicycles. As in the case of the work of Ptolemy, we shall briefly describe the geometrical constructions he employed. (p. 331)

As regards the motion of the earth round the sun Copernicus had of course nothing essential to add to the excentric circle (or concentric circle with an epicycle) which Ptolemy had used for the motion of the sun. He made the excentricity of the orbit equal to 0.0323 and the longitude of the apogee = 96° 40'2. Here again he did not make sufficient allowance for the inaccuracy of Greek and Arabian observations. (p. 331)

The motion of the moon was by Copernicus represented <u>by constructions much simpler</u> than those of Ptolemy. The equation of the centre he accounts for by an epicycle, but for the second inequality <u>he rejects the excentric deferent and uses instead a second epicycle</u>. (p. 333)

In the planetary theories Copernicus had the great advantage over Ptolemy, that he had (as regards the motion in longitude) <u>only the first inequality to deal with</u>, the period of which is the sidereal period of revolution. This <u>Ptolemy had accounted for</u> by the excentric circle and the equant or circle of uniform angular motion, the centre of the deferent or circle of equal distances being half-way between the earth and the centre of the equant. Copernicus <u>might have adopted this arrangement</u>, but he considered that the principle of uniform circular motion had been violated by the introduction of the equant, and <u>he had therefore to find some other explanation</u>. For the outer planets this was comparatively easy. In the figure, d is the <u>centre of the earth's orbit</u>, <u>to which point</u>, as representing the mean motion of the sun (i.e. of the earth) Copernicus <u>always referred the planetary motions.</u> The centre of the excentric orbit of the planet is at c, while the planet moves on an epicycle in the same direction and with the same angular velocity with which its centre moves round the excentric. The radius ae of the epicycle is one-third of the excentricity cd of the deferent, in fact $cd + ae$ is equal to Ptolemy's excentricity of the equant, so that instead of bisecting the excentricity as Ptolemy had done, Copernicus gave ¾ of it to the deferent and let the epicycle account for the rest, but <u>the result is the same</u>. (pp. 334-336)

In the Copernican system the <u>Ptolemaic epicycles of Venus and Mercury became the orbits of the two planets round the sun</u>. But the greatest elongations of these planets are not always equally great, a fact which is partly caused by the excentricity of their orbits, partly by that of the earth's orbit. In the case of Venus the phenomenon is simple enough, since her own orbit has a very small excentricity; and Copernicus therefore adopted a movable excentric after the manner of Apollonius, i.e. <u>he let the centre of the orbit of Venus move round the mean centre of the planet's orbit in a small circle</u> with twice the angular velocity of the earth and in the same direction. Whenever the earth passes the produced line of apsides of Venus at *a* and *b,* the centre of the excentric is at the point *m* of the small circle nearest to the mean sun, and the radius of the small circle is one-third of the average excentricity, $dn = 1/3$ cd ³ But owing to the very great excentricity of the orbit of Mercury (1/5, or more than twice that of Mars) <u>this theory was not sufficient for that planet</u>. (pp. 336-337)

The principal elements of the planetary orbits had been determined anew, and <u>though this was done on the basis of an utterly insufficient number of new observations,</u> <u>this was a defect which nobody seems to have remarked at the time. Another and a more serious defect</u>, partly caused by <u>the want of new observations</u>, partly <u>by an excessive confidence in the accuracy of Ptolemy's observations</u>, was that Copernicus in many cases <u>had kept too close to his great predecessor</u>. The man who had deposed the earth from its proud position as the centre of the universe and had recognized it to be merely one of the planets, had yet felt compelled to give it quite an exceptional position in his new system. Though he had said "in the midst of all stands the sun," <u>he had in his planetary theories assumed the centre of all movements to be the centre of the earth's orbit, where the sun was not</u>. And the year, that is to say the period of revolution of the earth, was intimately connected with the motion of the two inner planets (Mercury and Venus), both in longitude and latitude, and the same was the case with the motion of the outer planets (Mars, Jupiter and Saturn) in latitude, <u>so that the earth was nearly as important a body in the new system as in the old</u>. <u>Nor had the motion of the earth done much to simplify the old theories</u>, for though the objectionable equants had disappeared, <u>the system was still bristling with auxiliary circles</u>. (pp. 342-343)

The book of Copernicus was published in <u>1543</u>, and Kepler's book on Mars, showing that the planetary orbits are ellipses, appeared in <u>1609</u>. In the same year the telescope was first directed to the heavenly bodies and completely <u>changed many of the prevailing notions as to their constitution</u>. The period from 1543 to 1609 was a transition period, as the system of Copernicus had not yet been purified and strengthened by Kepler. (p. 345)

In the eighth chapter of his book on the comet of 1577 (where the parallax of Mars is not mentioned) Tycho describes his own system, which he says he had found "as if by inspiration" four years before the book was written, that is, in <u>1573</u>. The <u>earth</u> is the centre of the universe and the centre of the orbits of the moon and the sun, as well as of the sphere of the fixed stars, which latter revolves round it in twenty-four hours, carrying all the planets with it. The <u>sun</u> is the centre of the orbits of the five planets, of which Mercury and Venus move in orbits whose radii are smaller than that of the solar orbit, while the orbits of Mars, Jupiter, and Saturn encircle the earth. (p. 363)

<u>This system is in reality absolutely identical with the system of Copernicus, and all computations of the places of planets are the same for the two systems. As it leaves the earth at rest</u>, the Tychonic system might serve as a stepping-stone from the Ptolemaic to the Copernican system, and one might have expected it to have been proposed before the latter. (pp. 363-364)

<u>Unlike Copernicus, Tycho had at his disposal a great mass of observations</u>, made during many years according to a well-considered plan of following the sun, moon, and planets <u>right round the heavens and not merely observing them occasionally at opposition or other interesting points of their orbits. Hereby he succeeded in making the first important step forward since Ptolemy as regards the motion of the moon</u>, so that at his death all the great lunar perturbations were known with the one exception of the secular acceleration of the mean motion, which could only be discovered by the comparison of observations made in the course of centuries. <u>The motion in longitude he represented in a manner different from that of Copernicus and agreeing better with the observed positions</u>. (p. 368)

During the last year of his life, on the completion of his lunar theory, Tycho had commenced to investigate the motions of the planets, in which work Kepler became associated with him, but in October, 1601, Tycho's death set Kepler free <u>to prosecute the work in his own way</u>. (pp. 370-371)

Though Tycho had rejected the motion of the earth, he had, <u>mathematically speaking</u>, adopted the system of Copernicus, and by proving comets to be celestial bodies he had finally put an end to the idea of solid spheres, whereby he greatly increased the chance of success of the new system. In his writings Kepler repeatedly claims for Tycho the merit of having "destroyed the reality of the orbs". Another ancient error which Tycho practically abolished was the belief in the irregular motion of the equinoxes, <u>which he showed to have been caused solely by errors of observation</u>. (p. 371)

The great practical astronomer had indeed <u>thoroughly shown the insufficiency of all previous theories</u>, but <u>he had at the same time increased the accuracy of observed positions so vastly</u>, that it would <u>now be possible</u> to produce a satisfactory theory and, better still, <u>to determine the actual orbit in space in which each planet was travelling</u>, <u>a feat never yet attained</u>. The material for the investigation was ready, thanks to Tycho, and the mathematician to make use of it was also ready. (p. 372)

III

Pierre Maurice Marie Duhem

Pierre Duhem. *To Save the Phenomena – An Essay on the Idea of Physical Theory from Plato to Galileo*. Translated by Edmund Doland and Chaninah Maschler. University of Chicago Press. 1969. *Translated from the French text, originally published as* "ΣΩZEIN TA ΦAINOMENA: *Essai sur la notion de théorie physique de Platon à Galilée*," Annales de philosophie chrétienne,

79/156 (ser. 4, V1), 113–38, 277–302, 352–77, 482–514, 576–92. In 1908 it was reprinted under the same title by A. Hermann et Fils.

Pierre Duhem Ph. D. (1861–1916) was a French physicist and historian and philosopher of science. As a physicist he championed "energetics," holding generalized thermodynamics as foundational for physical theory, that is, thinking that all of chemistry and physics, including mechanics, electricity, and magnetism, should be derivable from thermodynamic first principles. In philosophy of science, he is best known for his work on the relation between theory and experiment, arguing that hypotheses are not straightforwardly refuted by experiment and that there are no crucial experiments in science. In history of science, he produced massive groundbreaking work in medieval science and defended a thesis of continuity between medieval and early modern science. He is regarded as having created the field of the history of medieval science. He is well known for his work on the history of science, which resulted in the ten volume *Le système du monde: histoire des doctrines cosmologiques de Platon à Copernic* (*The System of the World: A History of Cosmological Doctrines from Plato to Copernicus*). Pierre Duhem was a devout Catholic.

Quotations of Duhem

The condemnation carried through by the Holy Office resulted from the clash between two realist positions. This head-on collision might have been avoided, the debate between the Ptolemaists and Copernicans might have been kept to the terrain of astronomy, if certain sagacious precepts concerning the nature of scientific theories and the hypotheses on which they rest had been heeded. These precepts, first formulated by Posidonius, Ptolemy, Proclus and Simplicius, had, through an uninterrupted tradition, come down directly to Osiander, Reinhold and Melanchthon. But now they seemed quite forgotten.

There were, however, voices of authority to call attention to them once again. One of these was Cardinal Bellermine's, the same, who in 1616 was to examine the Copernican writings of Galileo and Foscarini. As early as April 12, 1615 Bellarmine had written Foscarini a letter full of wisdom and prudence. (pp. 106-107)

Galileo's notions of the validity of the experimental method and the art of using it are nearly those that Bacon was later to formulate. Galileo conceives of the proof of a hypothesis in imitation of the *reductio ad absurdum* proofs that are used in geometry. Experience, by convicting one system of error, confers certainty on its opposite. Experimental science advances by a series of dilemmas, each resolved by an *experimentum crucis*.

Since this manner of conceiving of the method of experiment was so simple, it was bound to become extremely fashionable; but because it was too simple, it was entirely in error. Grant that the phenomena are no longer saved by Ptolemy's system; the falsity of that system must then be acknowledged. But from this it does not by any means follow that the system of Copernicus is true; the latter is, after all, not purely and simply the contradictory of the Ptolemaic system. Grant that the hypotheses of Copernicus manage to save all the well known phenomena; that these hypotheses *may* be true is a warranted conclusion, not that they are *assuredly* true. Justification of this last proposition would require that one prove that no other set of hypotheses could possibly be conjured up that would do as well at saving the phenomena. The latter proof has never been given. Indeed, was it not possible, in Galileo's own time, to save all the appearances that could be mustered in favor of the Copernican system by the system of Tycho Brahe?

These logical observations had often been made before Galileo's time. Their justice struck the Greeks the day Hipparchus succeeded in saving the solar motion by either an eccentric or an epicycle. Thomas Aquinas had formulated them with the utmost clarity. (pp. 109-110)

Many philosophers since Giordano Bruno have taken Osiander harshly to task for the preface he placed at the head of Copernicus' book. And Cardinal Bellarmine's and Pope Urban the Eight's counsels to Galileo have been treated with hardly less severity since the day they were first published.

The physicists of our day, having gauged the worth of the hypotheses employed in physics and astronomy more minutely than did their predecessors, having seen so many illusions dissipated that previously passed for certainties, have been compelled to acknowledge and proclaim that logic sides with Osiander, Bellarmine and Urban VIII, not with Kepler and Galileo – that the former had understood the exact scope of the experimental method and that, in this respect, Kepler and Galileo were mistaken. (p. 113)

Table of Contents

Dedicated to St. Robert Bellarmine i

Quotations iii

Preface xvii

The Poupard "Solution" of the Galileo Case 1

Part One: Why I Mistrust Card. Paul Poupard 3

Part Two: "Faith Can Never Conflict With Reason" 13

Part Three: A List of Observations Regarding the Address by His Holiness Pope John Paul II On "The Galileo Case" 31

Appendix A: Part of Galileo's Letter to Kepler 63

Appendix B: Excerpts from the Letter of Bellarmine to Foscarini . 65

Epilogue 67

Preface

The present book fits philosophically and scientifically in our "Assertions and Refutations" series with "Imagination Misused" written by my friend and colleague Dr. Donald Boland as "Assertions and Refutations VII". In his book Dr. Boland subjects an article written by the American Catholic Physicist, Stephen Barr, to a thoroughgoing critique from the perspective of the natural philosophy of Aristotle and St. Thomas Aquinas.

In addition to that book, which concentrates on the modern materialistic theory of evolution as defended by Barr, Dr. Boland has contributed in large measure to the composition of the Preface of this present book: in content and style; and in ordering it to its end as a fitting introduction to "The Poupard 'Solution' of the Galileo Case" written by Frits Albers. For this excellent and much valued contribution, I acknowledge with gratitude my debt to him.

The figure of Paul Poupard, who was to become cardinal archbishop of Paris, together with holding other highly significant positions in the Catholic Church and the Institut Catholique, looms large over the pontificate of Pope Saint John Paul II, as may be seen simply from what is referred to in this book of Frits Albers.

Albers points to machinations in which such a high profile office-bearer of the Vatican curia was involved as context for the argument he makes for an interpretation of the speech made by Pope Saint John Paul II to the Pontifical Academy of Sciences in October 1992, in which the case of Galileo was a principal matter of discussion.

One of the faithful with implicit trust is his or her pastors (bishops) would find it hard to believe that there could be machinations, which Albers is prepared to describe as betrayals, within the heart of the Catholic Church hierarchy, let alone at the highest level. But even a superficial study of the history of the Church especially at the highest level would quickly disabuse him or her of such naivety. Albers makes the point that even God himself chose one who would betray him.

Every such person takes advantage of his position of trust and awaits his opportunity, which in this case was apparently the incapacitation of the pope for an extended period of time following the assassination attempt on his life on 13 May 1981. This date just happened to coincide with the hundredth anniversary of the birth of Teilhard de Chardin, of whom Poupard was evidently a devotee.

We will not go over the details Albers has provided. But it would perhaps be of some use to know something of the thinking of Poupard that would explain his motivation to act as he did. We leave it to the readers to judge its import in regard to the events that took place.

However, first let us say something about the "atmosphere" in the Vatican around that time. Pope Saint Paul VI had described it as affected by "the smoke of Satan". Poupard was not the only one scheming to divert the Church from its divine mission. We do not need to enter into the question of his subjective guilt.

Infiltrating a desire to bring it into the modern world after Vatican II, by using in particular the "philosophy" of Teilhard de Chardin, was only one way. Connected with this was the creation of the belief that the Church had failed to adjust its thinking to modern science, of which the case of Galileo was imagined to be solid evidence.

But we must not forget that there were various other "forces" bent upon destroying the influence of the Catholic Church by destroying the Church itself, not just intellectually/culturally, but in other ways such as directly politically. We will not go into these other kinds of machinations, though they are all linked in one way or another.

Frits Albers was focused on the mischief to be made by confusing the notions of science and philosophy and particularly with the role of Poupard in this. We may gain some insight into his motivation if we are able to study an example of the workings of his mind.

Often there is no opportunity given for such an exercise, but fortunately we have a report of a speech he gave in 2001, that enables us to gain some insight into his thinking and possible motivation for engaging in the machinations disclosed by Frits Albers.

The speech is one he gave at the University of Limerick in Ireland described as follows: "From Fear to the Beauty of Mystery" A Contribution to the discussion on The Relationship between Technology and Culture at Mary Immaculate College in the University of Limerick, Wednesday 27 June, 2001.

The reader can readily obtain access to the speech but we will draw attention to parts of it and make some comments that hopefully will support the case that Albers makes in this book of his.

First of all, the title is very suggestive. For Poupard, like Teilhard de Chardin, is clearly what may be described as a supreme optimist where modern science is concerned. It trumps everything - no need to worry about philosophy, especially that of the premodern Aquinas. Theology, well, we will adjust the doctrines of the Church where necessary.

The only thing we need to fear, as President Franklin Roosevelt said, is fear itself. Does that include fear of the Lord? Poupard adopts the favourite ploy of painting those who would point to something amiss with the march of progress fuelled by modern science as timid souls fearful of the glorious future promised in a brave new world. Join the bold and beautiful in the vanguard of Lenin, and charge into the night.

At the very outset he has said: "The basis for all I say is the firm conviction that there is no inner contradiction between the viewpoint of any scientist and that of men and women who believe in God." Perhaps, being in the British Isles, he had heard of Richard Dawkins, that famous propagator of the theory of evolution. What was the title of his famous book on the view scientists should have of belief in God?

There we have enough to see that Poupard is divorced from reality when he thinks of modern science and religious belief. We may immediately apply to him what Dietrich von Hilderbrand applies instantly to Teilhard de Chardin – "utter philosophical confusion", "theological primitiveness" and "crass naturalism". Looking into his earlier education we see that he "majored in history and theology" - he seemed to have skipped philosophy. So, as appears in this very speech, he is more familiar with such experts on modern culture as Martin Heidegger, Aldous Huxley and Nicholas Berdyaev. Aristotle and Aquinas apparently are too old fashioned (anti-modern science) to be taken seriously.

Thus, we have the toxic mixture of two extreme errors, naturalism based on modern science without philosophy, and supernaturalism (fideism) immersed in mystery that accommodates the irrational. He is not as vociferous in this regard as Teilhard de Chardin, but in a way this is more dangerously deceptive especially when we take his ecclesiastical position of authority into account.

At least Teilhard showed his true colours and proclaimed his erroneous "philosophy" of science and religion loudly and clearly. Poupard wished to give the impression that he was merely putting forward what the pope really thought. One wonders how it was possible for the pope not to see through him. Albers contention is that he did. But the situation had reached such a dire stage of complex deception in the world generally and "the smoke of Satan" had even entered into the inner circles of the Church that a direct exposure was not necessarily the most prudent approach. That is part of Albers' point. Have we reached the stage when even the elect might be deceived?

The quote below is typical of Poupard's use of the pope's words: "Suffice it to say that the present Pope made it clear from early on in his pontificate that 'the Church freely recognizes... that it has benefited from science'; on several occasions he has drawn attention to the fact that 'collaboration between religion and modern science is to the advantage of both, and in no way violates the autonomy of either'. (*Address to the Pontifical Academy of Sciences at the Commemoration of Albert Einstein, 10 November 1979*)."

The pope's measured giving of credit to modern science (and its material/empirical method) - which he specifically says is inadequate without philosophy (even natural philosophy) - is twisted to mean modern science and its method pure and simple. Why? Because Poupard and the curial crew around him do not have any sound philosophy to control their uncritical admiration for an imperfect science that is universally taken as the only kind of science there is.

That makes for a profound mistake and incorrigible deception. Albers brings this out from the very words of Pope Saint John Paul II himself. It may be noted that the pope does this in a language that is not strictly Aristotelian/Thomist but in one where he hopes the audience will get the message. Unfortunately, it seems that very few, if any, did, so imbued were they in the mistaken notion of science that not just rejects philosophy as Metaphysics but also as Natural Philosophy (as has been explained *ad nauseam* in our series). The pope therefore used the blanket term "meta-scientific" which obviously Poupard and Co. had not the slightest clue what it meant - except perhaps the "mysteries" of theology.

Poupard has to have no understanding of the pope's encyclical *Fides et ratio*. For it is written in the language of Aristotle and Aquinas. Who else uses such terms as "metaphysical" as the most meaningful in science and philosophy? Who else identifies modern philosophy as reversing the relation between being and knowledge. Instead of knowledge being a form of being, as is necessary to understand the concept of truth, the modern mind, as the pope says, thinks first in terms of human knowledge and puts being as but an object of (human) knowledge?

Yet, Poupard goes on to assert: "This conviction [that modern science as limited to knowledge of the physical universe is what is meant by "science" in St. Thomas] is at the heart of the Encyclical

Letter *Fides et Ratio* ... This is a conviction I wholeheartedly share, particularly since the days when the Holy Father asked me to work on the commission set up to re-evaluate the Galileo case ...". What was his state of mind before being appointed?

Immediately after making this declaration of sharing the pope's understanding of science, who does Poupard appeal to in order to throw light on the subject of the twin of modern science, modern technology, Martin Heidegger! We quote Poupard's enlightening conclusion on what Heidegger says on the question of technology: "It is never enough to understand technology in the sense of seeing how it works. It is more important, but more demanding, to discover what technology means. It is impossible to assess its effects on culture without this ability to step outside the immediate experience."

Cardinal Poupard goes on to discuss the fear of technology which is to be diagnosed as a fear of the new. As there is mystery at the heart of this modern phenomenon it is challenging, but the Christian, following the Vatican Council and Pope John Paul II's encouragement, should see what is positive in modern science and technology.

The speech ends with linking technology to poetry as Heidegger does. This turning, one supposes, is part of the mystery that is in the essence of technology. But more likely it is but a symptom of the confusion of mind that leads one with the original confident optimism regarding modern science in the end to wax lyrical about the future.

We have put these comments mainly to bring out the problem caused by the use of the term "science" in two opposed meanings.

The first is that of Aristotle and Aquinas that is what we might call perfect science which has its full meaning in Metaphysics (and Ethics) and the second is what we might call imperfect science, which limits the meaning not only to the physical or material order of reality but even there to the line of explanation that Aristotle called material cause.

The pope tends to use the word "science" in the second imperfect sense, as is the sense used by those who prepared the report. He is moreover constrained to do so given the audience. This is what deceives Poupard and others, who believe the pope is accepting the meaning of science as they understand it. But the pope makes it abundantly clear that science so taken needs to be supplemented by philosophy, and by this evidently means philosophy as understood by Aristotle and Aquinas, even to including Metaphysics.

Thus he requires that science be subject to meta-scientific principles. Frits Albers brings to bear this point of the reading of the report that is fatal to the likes of Poupard or anyone wanting to defend Teilhard de Chardin or Galileo's position on the nature of science.

But we leave the details of the latter for Albers to explain. The main point of his book is to show up the interpretation of the pope's reading of the report as Poupard and others would have us believe.

Further to supplement the present book we refer readers to the author's detailed analysis of the works of Teilhard de Chardin in the following two books: "Teilhard de Chardin and the Dutch Catechism" and "The 'Theology' of the Late Pierre Teilhard de Chardin S.J."; both of which are published in *The Selected Works of Frits Albers, Volume One*, subtitled "Analysing the Errors and Exposing the Real Agenda of Pierre Teilhard de Chardin S.J."

These books of Frits Albers were in fact written early in the tumultuous period following the Council and show a remarkable depth of insight into the systematic modernism of de Chardin and its devastating impact on post–conciliar Catholic life.

In analysing the fundamental philosophic principle of the works of Teilhard, Albers resembles the anatomist who traces out the real structure of the literary corpus he is dissecting to lay bare its bones, sinews and organs for all to see in plain daylight. The autopsy is not a pretty sight but is one that must be faced and its consequences understood; for it is true to say that the modern theory of evolution, especially as proposed by Teilhard with his calculated assertions against Catholic Faith and Dogmas of the Church, is the philosophic fulcrum around which turns the formation (deformation) of many Catholics educated in modern science and theology. Those so educated will have missed the philosophical distinctions made by Pope St. John Paul II when he read out the Address handed him by Cardinal Poupard as the 'Solution' of the Galileo case.

As an aid better to understand the necessary and basic distinctions between philosophy and modern science as used by the pope; and the impotence of modern science (e.g. the hypotheses and geometrical constructions of astronomy) inherent in its own nature to define itself and to assign to itself its rightful place in the hierarchy of human knowledge (to say nothing of theological knowledge), a list of Quotations is provided at the head of this book. The reader is invited to ponder these Quotations and to decide for him or herself if Albers has employed these principles and distinctions and understood their import in what he has to say concerning the Poupard "Solution" of the Galileo case.

—Frank Calneggia
Perth, Australia

The Poupard "Solution" of the Galileo Case

In line with the subtitle on the title page of this documentary we may say:

**A List of Betrayals by One Cardinal,
Paul Poupard, Archbishop of Paris,
which neither deceived the Holy Father
nor were used by him to deceive**

As far as the main title of this paper is concerned: the essentials of the Galileo Case were resolved by the Holy Catholic Church (which means with the direct assistance and under the direct guidance of the Holy Spirit) in the seventeenth century, when the wonderful gifts and a unique combination of talents of a great Saint, *St. Robert Bellarmine,* were at the disposal of our Holy Mother the Church. The Galileo Case has been resolved once and for all by authority, and never by the driftsand of what nowadays goes under the name *culture,* or *rationalist insights* or, what is the same, *new theology,* and least of all by *science.* All these, according to a brilliant book by the late Professor *Richard Weaver* of Chicago, *Ideas have Consequences,* were merely the witches which kept the poisonous brew boiling that had been concocted strictly according to the nominalist recipe of one *William of Occam,* the 'father of atheism'. For it had been this medieval dissenter and excommunicated fugitive from papal jurisdiction, William of Occam, who, for millions of his Renaissance and subsequent followers right up into our own times, had severed irrevocably Faith from Reason, Truth from truth.

It proved to be the same fatal cut for Cardinal Paul Poupard and his entourage of innumerable existentialists and teilhardians....

This brings us to the sub-title of the present paper.

History has recorded several major betrayals by Cardinal Paul Poupard, Archbishop of Paris and president of its Institut Catholique. I will briefly describe two of them here as an introduction to his major one, *his 'resolution' of the Galileo Case.*

Part One

Why I Mistrust Card. Paul Poupard

[Excerpts taken from the book
The Jesuits by Malachi Martin]

The setting is the papal meeting of Pope John Paul II with 6 of his most powerful cardinals: 'Dottrina', 'Propaganda', 'Clero', 'Vescovi' 'Religiosi' and 'Stato', in the middle of the Northern Spring, 1981.

"In sum, *Propaganda* agreed both with the 1962 condemnation of the Jesuit de Chardin, and with the Holy Father's indictment of the society as a whole in 1981".

"It seemed at first as if *Clero* would confine his contribution to an amplification of *Propaganda's* link between Teilhard de Chardin's work and present-day Jesuit activity. Why was it, he seemed merely to muse about the problem a bit further, that the Jesuit faculties of philosophy and theology at the Sèvres Centre in Paris were organising a celebration for the coming June 13 to honour the centenary of de Chardin's birth. According to *Clero's* information, they were doing so with the blessing of Pontifical Institutes in Rome and the approval of the Secretariat of State and of the Jesuit General. His Holiness's suggestion was more pointed. The Pontiff was sure that *Stato* would communicate to Father Arrupe the Holy See's disapproval of the planned celebration". (p. 89).

"When the Holy Father was shot, *Stato*, on a formal visit to the United States at the time, hurried back home to take control of the

Holy See as Vatican Secretary of State. In those hectic, suspicion-laden days of May and June of 1981, there was no medical certainty that the Pontiff would pull through. It would, as it turned out, take the Holy Father the best part of six months to get back to anything like a full schedule. In hindsight, many are forced to the conclusion that there were those, including both *Stato* and Arrupe, who considered that John Paul's grip on papal affairs had been loosened once and for all. They did not expect him to recover, to get back into harness. That is the most obvious reading of *Stato's* and Arrupe's behaviour in the immediate aftermath of the May 13 shooting.

One of *Stato's* first public acts on his return was a direct violation of John Paul's will expressed at the papal meeting: He sent a highly congratulatory message to *Archbishop Paul Poupard*, President of the Secretariat for Non-Believers, lauding the work and thought of Father Teilhard de Chardin, whose centenary the 'Institut Catholique' of Paris was celebrating. *Stato's* message praised 'the amazing echo of his (de Chardin's) research, joined with the radiance of his thought', all of which 'has left a durable mark on his age'.

It was an enormous gaffe of disproportionate arrogance. And although *Stato* dated the message May 12 - one day before John Paul was shot - clearly it was written and sent after the event.

"Arrupe followed suit almost immediately with what seemed a calculated and feckless disregard for John Paul's opinions and bidding. He sent a message dated May 30, and went even farther than *Stato* in his praise of de Chardin"... (p. 94-95).

"By the time John Paul II had recovered sufficient energy and his doctors allowed him some activity, toward the latter half of July 1981, the decision to remove Arrupe by hook or by crook had been made by the Jesuit's accumulated enemies in the Vatican Curia and in the Latin American Church. Almost certainly *Vescovi, Dottrina, Propaganda, Clero*, powerful Latin American churchmen such as Archbishop Alonzo Lopez Trujillo of Medellin, Colombia, and some older Jesuits of a conservative, anti-Arrupe bent were in on that decision. Arrupe had to go.

John Paul II acquiesced readily. In fact, when he learned how *Stato* and Arrupe had been behaving, the Pope's own reaction was visceral. As an added sting to his reaction, he decided not to inform *Religiosi* of his papal decision to remove Arrupe. This was tantamount to insult: *Religiosi* was the cardinal directly responsible for the behaviour of all Religious priests and of Arrupe in particular. Since the shooting, John Paul had wanted nothing of that cardinal in his life".

"*Stato* and Arrupe were the Pope's targets, however. Quickly, he hit *Stato* with a typically Roman punishment for his transgressions. The Press Office of the Holy See and the official Vatican newspaper, *Osservatore Romano*, both private stamping grounds of *Stato's* were forced by papal order to publish an official statement correcting *Stato's* praise of de Chardin and repeating the condemnation of 1962. The put-down was public". (p. 96 – 97).

Here follows the official text of this "public put-down", issued by the Holy See press office on July 11, 1981, as it appeared in the *Osservatore Romano* of July 20 1981, mentioning Archbishop (by then not yet Cardinal) Paul Poupard by name:

"The letter sent by the Cardinal Secretary of State to His Excellency Mons. Poupard on the occasion of the centenary of the birth of Fr. Teilhard de Chardin has been interpreted in a certain section of the press as a revision of previous stands taken by the Holy See in regard to this author, and in particular of the *Monitum* of the Holy Office of 30 June 1962, which pointed out that the work of the author contained 'ambiguities and grave doctrinal errors'.

The question has been asked whether such an interpretation is well founded. After having consulted the Cardinal Secretary of State and the Cardinal Prefect of the Sacred Congregation of the Faith, which, by order of the Holy Father, had been duly consulted beforehand about the letter in question, we are in a position to reply in the negative.

Far from being a revision of the previous stands of the Holy See, Cardinal Casaroli's letter expresses reservations in various passages - and these reservations have been passed over in silence by certain newspapers - reservations which refer precisely to the judgment given in the Monitum of June 1962, even though this document is not explicitly mentioned."

The second example of betrayal involving the *Institut Catholique* of Paris and its president, Archbishop Paul Poupard, centres on interference with the dissemination of truth by means of the direct and wilful suppression of Catholic scholarship in favour of a free and unencumbered promotion of doctrinal errors. The scandal appeared in the March 19, 1992 edition of the American

Catholic paper *The Wanderer,* quoting two other major European Catholic periodicals, *30 Days* and *Il Sabato.*

Reporting a scandal: An editorial in the current issue of *30 Days* magazine (issue no. 2), titled "Scandal at the Institut Catholique" raises some tough questions about the openness of modern biblical scholars to research which offers evidence that the Gospels were written by A.D. 50.

Reporting on investigative work conducted by the Italian Catholic weekly *Il Sabato* the editorial asks why the Institut Catholique in Paris will not allow to be printed, or even acknowledge the existence of, the biblical scholarship of *Fr. Jean Carmignac.*

Fr. Carmignac, until his death in 1986, was one of the world's leading experts in Hebrew and Aramaic, and his extensive research in language and the Fathers of the Church led him to believe Matthew, Mark, and Luke had written their Gospels by A.D. 50. In addition, Carmignac noted the scholarship of 49 other recognised experts who agreed with him, but whose works also had either been ignored or censored or else they did not dare wage a battle in the name of their scientific conviction."

"For the consequences", stated the *30 Days* editorial, "would have revolutionised the dominant exegetical trends today. Many ideas, whose certainty is taken for granted today, would have crumbled ... If the Synoptic Gospels were written in a Semitic language it means they were written soon after Jesus' years on earth, when the protagonists were still alive. It means that the Synoptic Gospels are the testimonies of people who saw and heard, of witnesses to the facts. It means they are not late elaborations by anonymous transcribers of popular traditions."

In 1983. Fr. Carmignac published a small book containing his findings and conclusions, and promised a later book which he described as "more convincing than ever and, I hope, irrefutable."

"But at that time an effort began to bury his work, the editorial said, under hefty shovelfuls of earth ..." Six years after his death, none of these texts has ever been published. An impenetrable curtain of silence has fallen on Fr. Carmignac and his work. The Catholic weekly *Il Sabato* has been hunting down his manuscripts. It discovered that Fr. Carmignac's entire archive is to be found at the Institut Catholique in Paris where he had taught. In all these years, the Institut Catholique has taken care not to tend to the publication of those pre-announced works, and, above all, it has prohibited people from seeing the material when they ask to see it ..."

One of the 49 scholars mentioned here by the late Fr. Jean Carmignac is no doubt *Claude Tresmontant*, whose magnificent book on that very same topic, *The Hebrew Christ* carries a lengthy foreword by the Most Reverend Jean Charles Thomas, Bishop of Ajaccio, dated May 1, 1983, three years before the death of Fr. Jean Carmignac. In his Foreword bishop Thomas refers specifically to the same general state of affairs as was reported by the three Catholic papers mentioned above. There is no change of heart in either the 'Institut Catholique' or its president, Paul Poupard ...

After this short introduction of the main historical characters in this critical dossier, and the indelible marks their lives have left on the pages of recorded history: a Renaissance mathematician,

Galileo Galilei; his most famous contemporary, a Renaissance Saint and God's chosen gladiator in the uneven contest, St. Robert Bellarmine; and the present-day teilhardian president of the corrupted 'institut catholique', Paul Poupard, Archbishop of Paris, we finally mention the one reality that is in no need of any introduction: *the Papacy of the Holy Roman Catholic Church*, Its roots go back more than 1600 years before Galileo, nearly 2000 years before Archbishop Poupard, and its infallible truths are still with us today, and will still be the most important part of recorded history when nothing else that human history could throw at it will matter anymore, being just as futile as it always has been.

According to an address by the Holy Father Pope John Paul II, presented to the Papal Academy of Sciences on October 31, 1992, this is how the various names found their way into this present record.

"I was moved by similar concerns on 10 November 1979 ... when I expressed the hope before this same Academy that 'theologians, scholars and historians, animated by a spirit of sincere collaboration, will study the Galileo case more deeply and, in frank recognition of wrongs from whatever side they come, dispel the mistrust that still opposes, in many minds, a fruitful concord between science and faith'.

A *Study Commission* was constituted for this purpose on 3 July 1981. The Commission is presenting today, at the conclusion of its work, a number of publications which I value highly. I would like to express my sincere gratitude to Cardinal Poupard, who was *entrusted* (my stress, because of the word *trust* used here) with coordinating the Commission's research in its concluding phase. To all the experts who in any way took part in the proceedings of the four groups that guided this multidisciplinary study, I express my profound satisfaction and my deep gratitude. *In the future, it will be impossible to ignore the commission's conclusions"*. (My stress of these truly prophetic words).

If we realise that the convening of this 'study commission' and the press release of the Holy Father's rebuff of Cardinal Casaroli's congratulatory letter to Archbishop Poupard were within a week of each other, showing that, at the time of the convening of the study commission, the Holy Father was fully aware of Archbishop Paul Poupard's duplicity, why then, we may well ask, did Pope John Paul II eventually appoint Paul Poupard as the coordinating spirit at the head of such a sensitive commission?

The answer, we think, has a lot to do with the reason why we read in the Gospel that the Apostle Judas was appointed by Our Blessed Lord to carry the common purse. If both had a genuine conversion, they would make a brilliant job of it. And if not? Then Paul Poupard would use the Pope's commission as Judas used the common purse: for his own private use and advancement. From which it would follow inevitably, that the Holy Father would be presented with a mongrel of a report. In both cases not much damage could be done. In Judas' case it would only be money, *"that tainted thing"*, and in Poupard's case it would merely amount to easily corrected corruptions of Papal teachings, of historical facts and of scientific data, proposed for the pursuit of his own ends. And just as Our Lord openly declared: *"One of you will betray Me"*, so the papal speech of October 31, 1992, would, if there was no real conversion on the part of Cardinal Poupard, convey the same message to the whole Church.

If now we couple the Fr. Jean Carmignac episode of the early '80s with *their recording* in the Press in 1992, then it becomes obvious not only that things did not start to look too good for a genuine conversion of Paul Poupard in the early and middle '80s, but that, in 1992, there still was no sign of it.

For, while Paul Poupard addressed the Holy Father in Rome as late as October 31, 1992, he was still actively suppressing the truth in the 'Institut Catholique' in Paris. So, as things turned out, the Holy Father was indeed presented in 1992 with a mongrel of a report from an unrepentant Cardinal Poupard and his equally unrepentant 'Galileo Commission'.

Since, going by his subsequent behaviour, Pope John Paul II had apparently made up his mind to read out to the Church whatever report he would be given by Card. Poupard, it is of the utmost importance that this report is read with the greatest of care. With that in mind I made a verbatim copy of it line by line, making sure that each sentence received its appropriate line number. To facilitate the discussion so that the unfolding of Cardinal Paul Poupard's duplicity can be followed more easily, and so can be clearly established, here follows first and foremost a reprint of the numbered line version of Pope John Paul's address of October 31, 1992.

This papal address appeared in the weekly English edition of the *Osservatore Romano* of November 4, 1992. The paper divided the entire speech into a kind of introduction (lines 1-5) and three main sections, indicated by the use of the Roman numerals I-III, (which are preserved in our accompanying line-by-line reprint), as well as into 14 subsections, numbered 1-14, which in our reprint are represented by the capital letters A-N because of our use of Arabic numerals for line numbering.

This address by the Holy Father was preceded by an address to the Pope by Cardinal Poupard. In it the Cardinal states:

> " ... and you had entrusted to *Cardinal Garrone* responsibility for coordinating the research ..."

But in line 25 of his address, the Holy Father contradicted this by stating:

> "I would like to express my sincere gratitude to Cardinal Poupard, who was entrusted with coordinating the Commission's research *in its concluding phase.*"

We all know it is the end result and the end report that count ... and these the Holy Father had **not** entrusted to Card. Garrone, but to no other than to Card. Poupard, although the latter did not wish to be publicly identified with it for obvious reasons of *trust* as will come to light in this analysis.

Part Two

"Faith Can Never Conflict With Reason"

A line-by-line representation of the address given by His Holiness
Pope John Paul II
to the members of the
Pontifical Academy of Sciences
on October 31, 1992.

From the
Osservatore Romano, November 4, 1992.

(I) A

1 The conclusion of the plenary session of the Pontifical Academy of Sciences gives me the pleasant opportunity to meet its illustrious members, in the presence of my principal collaborators and the Heads of the Diplomatic Missions accredited to the Holy See. To all of you I offer a warm welcome.

2 My thoughts go at this moment to Professor Marini-Bettolo, who is prevented by illness from being among us, and, assuring him of my prayers, I express fervent good wishes for his restoration to health.

3 I would also like to greet the members taking their seats for the first time in this Academy; I thank them for having brought to your work the contribution of their lofty qualifications.

4 In addition, it is a pleasure for me to note the presence of Professor Adi Shamir, of the Weizmann Institute of Science at Rehovot, Israel, holder of the Gold Medal of Pius XI, awarded by the Academy, and to offer him my cordial congratulations.

5 Two subjects in particular occupy our attention today. They have just been ably presented to us, and I would like to express my gratitude to Cardinal Paul Poupard and Fr. George Coyne for having done so.

B

6 In the first place. I wish to congratulate the Pontifical Academy of Sciences for having chosen to deal, in its plenary session, with a problem of great importance and great relevance today: *the problem of the emergence of complexity in mathematics, physics, chemistry and biology.*

7 The emergence of the subject of complexity probably marks in the history of the natural sciences a stage as important as the stage which bears relation to the name of Galileo, *when a univocal model of order seemed to be obvious.*

8 Complexity indicates precisely that, in order to account for the rich variety of reality, we must have recourse to a number of different models.

9 This realisation poses a question which concerns scientists, philosophers and theologians: how are we to reconcile the explanation of the world, beginning with the level of elementary entities and phenomena with the recognition of the fact that *"the whole is more than the sum of its parts"*?

10 In his effort to establish a rigorous description and formalisation of the data of experience, the scientist is led to have recourse to meta-scientific concepts, the use of which is, as it were, demanded by the logic of his procedure.

11 It is useful to state exactly the nature of these concepts in order to avoid proceeding to undue extrapolations which link strictly scientific discoveries to a vision of the world or to ideological or philosophical affirmations, which are in no way corollaries of it.

12 Here one sees the importance of philosophy which considers phenomena just as much as their interpretation.

C

13 Let us think, for example, of the working out of new theories at the scientific level in order to take account of the emergence of living beings.

14 In a correct method, one could not interpret them immediately in the framework exclusive to science.

15 In particular, when it is a question of the living being which is man, and of his brain, it cannot be said that these theories of themselves constitute an affirmation or a denial of the spiritual soul, or that they provide a proof of the doctrine of creation, or that, on the contrary, they render it useless.

16 A further work of interpretation is needed.

17 This is precisely the object of philosophy, which is the study of the global meaning of the data of experience and therefore also of the phenomena gathered and analysed by the sciences.

18 Contemporary culture demands a constant effort to synthesise knowledge and to integrate learning.

19 Of course, the successes which we see are due to the specialisation of research.

20 But unless this is balanced by a reflection concerned with articulating the various branches of knowledge there is a great risk that we shall have a "shattered culture", which would in fact he the negation of true culture.

21 A true culture cannot be conceived of without humanism and wisdom.

(II) D

22 I was moved by similar concerns on 10 November 1979, at the time of the first centenary of the birth of Albert Einstein, when I expressed the hope before this same Academy that "theologians, scholars and historians, animated by a spirit of sincere collaboration, will study the Galileo case more deeply and, in frank recognition of wrongs <u>from whatever side they come</u> (my stress), dispel the mistrust that still opposes in many minds, a fruitful concord between science and faith".[1]

23 A Study Commission was constituted for this purpose on July 3, 1981.

24 The very year when we are celebrating the 350th anniversary of Galileo's death, the Commission is presenting today, at the conclusion of its work, a number of publications which I value highly.

25 I would like to express my sincere gratitude to Cardinal Poupard, who was entrusted with coordinating the Commission's research in its concluding phase.

26 To all the experts who in any way took part in the proceedings of the four groups that guided this multidisciplinary study, I express my profound satisfaction and my deep gratitude.

---

[1] ABAS 71 (1979), pp. 1464-1465.

27 The work that has been carried out for more than 10 years responds to a guideline suggested by the Second Vatican Council and enables us to shed more light on several important aspects of the question.

28 In the future it will be impossible to ignore the Commission's conclusions.

29 One might perhaps be surprised that, at the end of the Academy's study week on the theme of the emergence of complexity in the various sciences, I am returning to the Galileo case.

30 Has not this case long been shelved and have not the errors committed been recognised?

31 That is certainly true.

32 However, the underlying problems of this case concern both the nature of science and the message of faith.

33 It is therefore not to be excluded that one day we shall find ourselves in a similar situation, one which will require both sides to have an informed awareness of the field and of the limits of their own competencies.

34 The approach provided by the theme of complexity could provide an illustration of this.

E

35 A *twofold question* is at the heart of the debate of which Galileo was the centre.

36 The *first* is of the epistemological order and concerns *biblical hermeneutics*.

37 In this regard, two points must again be raised.

38 In the first place, like most of his adversaries, Galileo made no distinction between the scientific approach to natural phenom-

ena, and a reflection on the nature of the philosophical order, which that approach generally calls for.

39 That is why he rejected the suggestion made to him to present the Copernican system as a hypothesis, inasmuch as it had not been confirmed by irrefutable proof.

40 Such therefore was an exigency of the experimental method of which he was the inspired founder.

41 Secondly, the geocentric representation of the world was commonly admitted in the culture of the time as fully agreeing with the teaching of the Bible, of which certain expressions, taken literally, seemed to affirm geocentrism.

42 The problem posed by theologians of that age was, therefore, that of the compatibility between heliocentrism and Scripture.

43 Thus the new science, with its methods and the freedom of research which they implied, obliged theologians to examine their own criteria of scriptural interpretation.

44 Most of them did not know how to do so.

45 Paradoxically, Galileo, a sincere believer, showed himself to be more perceptive in this regard than the theologians who opposed him.

46 "If Scripture cannot err", he wrote to Benedetto Castelli, "certain of its interpreters and commentators can and do so in many ways" [2].

47 We also know of his letter to Christine de Lorraine (1615) which is like a short treatise on biblical hermeneutics [3].

---

[2] Letter of 21 November 1613, in Adhesion nazionale delle Opere di Galileo Galilei dir. A. Favaro, edition of 1968, vol. V, p. 307 - 348.

F

48 From this we can now draw our first conclusion.

49 The birth of a new way of approaching the study of natural phenomena *demands a clarification on the part of all disciplines of knowledge.*

50 It obliges them to define more clearly their own field, their approach, their methods, as well as the precise import of their conclusions.

51 In other words, this new way requires each discipline to become more rigorously aware of its own nature.

52 The upset caused by the Copernican system thus demanded epistemological reflection on the biblical sciences, an effort which later would produce abundant fruit in modern exegetical works and which has found sanction and a new stimulus in the Dogmatic Constitution *Dei Verbum* of the Second Vatican Council.

G

53 The crisis that I have just recalled is not the only factor to have had repercussions on biblical interpretation.

54 Here we are concerned with the *second aspect* of the problem, its *pastoral dimension.*

55 By virtue of its own mission, the Church has the duty to be attentive to the pastoral consequences of her teaching.

---

[3] Letter to Christine de Lorraine, 1615 in Adhesion nazionale delle Opere di Galileo Galilei, dir. A. Favaro, edition of 1968, vol. V, pp. 307-348.

56 Before all else, let it be clear that this teaching must correspond to the truth.

57 But it is a question of knowing how to judge a new scientific datum when it seems to contradict the truths of faith.

58 The pastoral judgement which the Copernican theory required was difficult to make, in so far as geocentrism seemed to be a part of scriptural teaching itself.

59 It would have been necessary all at once to overcome habits of thought and to devise a way of teaching capable of enlightening the people of God.

60 Let us say, in a general way, that the pastor ought to show a genuine boldness, avoiding the double trap of a hesitant attitude and of hasty judgement, both of which can cause considerable harm.

H

61 Another crisis, similar to the one we are speaking of can be mentioned here.

62 In the last century and at the beginning of our own, advances in the historical sciences made it possible to acquire a new understanding of the Bible and of the biblical world.

63 The rationalist context in which these data were most often presented seemed to make them dangerous to the Christian faith.

64 Certain people, in their concern to defend the faith, thought it necessary to reject firmly-based historical conclusions.

65 That was a hasty and unhappy decision.

66 The work of a pioneer like Fr. Lagrange was able to make the necessary discernment on the basis of dependable criteria.

67 It is necessary to repeat here what I said above.

68 It is a duty for theologians to keep themselves regularly informed of scientific advances in order to examine, if such be necessary, whether or not there are reasons for taking them into account in their reflection or for introducing changes in their teaching.

I

69 If contemporary culture is marked by a tendency to scientism, the cultural horizon of Galileo's age was uniform and carried the imprint of a particular philosophical formation.

70 This unitary character of culture, which in itself is positive and desirable even in our own day, was one of the reasons for Galileo's condemnation.

71 The majority of theologians did not recognise the formal distinction between Sacred Scripture and its interpretation, and this led them unduly to transpose into the realm of the doctrine of the faith a question which in fact pertained to scientific investigation.

72 In fact, as Cardinal Poupard has recalled, Robert Bellarmine, who had seen what was truly at stake in the debate, personally felt that, in the face of possible scientific proofs that the earth orbited round the sun, one should "interpret with great circumspection" every biblical passage which seems to affirm that the earth is immobile and "say that we do not understand, rather than affirm that what has been demonstrated is false" [4]. [See Appendix B]

---

[4] Letter to Fr. A. Foscarini, 12 April 1615, cf. Adhesion nazionale delle Opere di Galileo Galilei, dir. A. Favaro vol. XII, p. 172.

73 Before Bellarmine, this same wisdom and same respect for the divine Word guided St. Augustine when he wrote:

"*If it happens that the authority of Sacred Scripture is set in opposition to clear and certain reasoning, this must mean that the person who interprets Scripture does not understand it correctly. It is not the meaning of Scripture which is opposed to the truth but the meaning which he has wanted to give to it. That which is opposed to Scripture is not what is in Scripture but what he has placed there himself, believing that this is what Scripture meant.*" [5]

74 A century ago, Pope Leo XIII echoed this advice in his Encyclical *Providentissimus Deus*:

"*Truth cannot contradict truth, and we may be sure that some mistake has been made either in the interpretation of the sacred words, or in the polemical discussion itself.*" [6]

75 Cardinal Poupard has also reminded us that the sentence of 1633 was not irreformable, and that the debate, which had not ceased to evolve thereafter, was closed in 1820 with the imprimatur given to the work of Canon Settele. [7]

J

76 From the beginning of the Age of Enlightenment down to our own day the Galileo case has been a sort of "myth", in which the image fabricated out of the events was quite far removed from reality.

---

[5] Saint Augustine, Epistula 143, n. 7; PL, 33, col. 588.

[6] Leonis X111 Pont. Max. Acta, vol. XIII (1894). p. 361.

[7] Cf. Pontificia Academia Scientiarum, Copernico, Galilei e la Chiesa. Fine della controversia (1820). Gli atti del Sant Ufficio, a cura di W. Brandmueller e E. J. Griepl, Firenze, Olschki, 1992.

77 In this perspective, the Galileo case was the symbol of the Church's supposed rejection of scientific progress, or of "dogmatic" obscurantism opposed to the free search for truth.

78 This myth has played a considerable cultural role.

79 It has helped to anchor a number of scientists of good faith in the idea that there was an incompatibility between the spirit of science and its rules of research on the one hand and the Christian faith on the other.

80 A tragic mutual incomprehension has been interpreted as the reflection of a fundamental opposition between science and faith.

81 The clarifications furnished by recent historical studies enable us to state that this sad misunderstanding now belongs to the past.

K

82 From the Galileo affair we can learn a lesson which remains valid in relation to similar situations which occur today and which may occur in the future.

83 In Galileo's time, to depict the world as lacking an absolute physical reference point was, so to speak, inconceivable.

84 And since the cosmos, as it was then known, was contained within the solar system alone, this reference point could only be situated in the earth or in the sun.

85 Today, after Einstein and within the perspective of contemporary cosmology, neither of these two reference points has the importance they once had.

86 This observation, it goes without saying, is not directed against the validity of Galileo's position in the debate; it is only meant to

show that often, beyond two partial and contrasting perceptions, there exists a wider perception which includes them and goes beyond both of them.

L

87 Another lesson which we can draw is that the different branches of knowledge call for different methods.

88 Thanks to his intuition as a brilliant physicist and by relying on different arguments, Galileo, who practically invented the experimental method, understood why only the sun could function as the centre of the world, as it was then known, that is to say, as a planetary system.

89 The error of the theologians of the time, when they maintained the centrality of the earth, was to think that our understanding of the physical world's structure was, in some way, imposed by the literal sense of Sacred Scripture.

90 Let us recall the celebrated saying attributed to Baronius: "Spiritui Sancto mentem fuisse nos docere quomodo ad coelum eatur, non quomodo coelum gradiatur".

91 In fact, the Bible does not concern itself with the details of the physical world, the understanding of which is the competence of human experience and reasoning.

92 There exist two realms of knowledge, one which has its source in Revelation and one which reason can discover by its own power.

93 To the latter belong especially the experimental sciences and philosophy.

94 The distinction between the two realms of knowledge ought not to be understood as opposition.

95 The two realms are not altogether foreign to each other - they have points of contact.

96 The methodologies proper to each make it possible to bring out different aspects of reality.

(III) M

97 Your Academy conducts its work with this outlook.

98 Its principal task is to promote the advancement of knowledge, with respect for the legitimate freedom of science [8] which the Apostolic See expressly acknowledges in the statutes of your institution

99 What is important in a scientific or philosophic theory is above all that it should be true or, at least, seriously and solidly grounded.

100 And the purpose of your Academy is precisely to discern and to make known, in the present state of science and within its proper limits, what can be regarded as an acquired truth or at least as enjoying such a degree of probability that it would be imprudent and unreasonable to reject it.

101 In this way unnecessary conflicts can be avoided.

102 The seriousness of scientific knowledge will thus be the best contribution that the Academy can make to the exact formulation and solution of the serious problems to which the Church, by virtue of her specific mission, is obliged to pay close attention - problems no longer related merely to astronomy, physics and

---

[8] Cf. Second Vatican Ecumenical Council, Pastoral Constitution *Gaudium el spes,* n. 36, par. 2.

mathematics, but also to the relatively new disciplines such as biology and biogenetics.

103 Many recent scientific discoveries and their possible applications affect man more directly than ever before, his thought and action, to the point of seeming to threaten the very basis of what is human.

N

104 Humanity has before it two modes of development.

105 The first involves culture, scientific research and technology, that is to say whatever falls within the horizontal aspect of man and creation, which is growing at an impressive rate.

106 In order that this progress should not remain completely external to man, it presupposes a simultaneous raising of conscience, as well as its actuation.

107 The second mode of development involves what is deepest in the human being, when transcending the world and transcending him - man turns to the One who is the Creator of all.

108 It is only this vertical direction which can give full meaning to man's being and action, because it situates him in relation to his origin and his end.

109 In this twofold direction, horizontal and vertical, man realises himself fully as a spiritual being and as 'homo sapiens'.

110 But we see that development is not uniform and linear, and that progress is not always well ordered.

111 This reveals the disorder which affects the human condition.

112 The scientist who is conscious of this twofold development and takes it into account contributes to the restoration of harmony.

113 Those who engage in scientific and technological research admit as the premise of its progress, that the world is not a chaos but a "cosmos": that is to say that there exist order and natural laws which can be grasped and examined, and which, for this reason, have a certain affinity with the spirit.

114 Einstein used to say: "What is eternally incomprehensible in the world is that it is comprehensible".[9]

115 This intelligibility, attested to by the marvellous discoveries of science and technology, leads us, in the last analysis, to that transcendent and primordial Thought imprinted on all things.

116 Ladies and gentlemen, in concluding these remarks, I express my best wishes that your research and reflection will help to give our contemporaries useful directions for building a harmonious society in a world more respectful of what is human.

117 I thank you for the service you render to the Holy See, and I ask God to fill you with his gifts.

In the next section (Part Three) we will come to grips with the interplay between a disloyal Cardinal and his 'report' and the Holy Father and his response to this report. For this we make the following **Preliminary Remarks.**

Catholics can only deal with the Papacy in the supernatural Light of Catholic Faith. In that Light we see clearly that the Papacy has only one exclusive use: <u>to be used solely by One</u>.

1. So there is only <u>one use</u>: the dissemination of Truth, the exposition of error, for the eternal salvation of souls.

---

[9] In *The Journal of the Franklin Institute*, vol. 221, n. 3, March 1936.

2. And there is only <u>One User</u>: the Holy Spirit through the medium of His One, Holy, Catholic and Apostolic Church.

Thus, if a group of manipulators tries to use the Papacy for its own ends: the dissemination of error and betrayal, then they will run up against the *Rock of Peter* and be crushed by it in the process. For the Only One who has exclusive use of the Papacy will continue to use it single-mindedly for His own exclusive use: to teach the Truth and expose error and betrayal, making sure that, at the same time, the whole satanic plot backfires on the conspirators.

Furthermore, if in the long history of the Papacy a Pope has taught a certain truth, then we not only know that this truth is *believed* by every other Pope, we know that it is *proclaimed* by every other Pope. The name of the Pope has become immaterial; what matters is that his name is **Peter**.

Thus, if Pope Pius XII teaches that science is unable by its own methods even to understand its own nature, and consequently is unable to allocate its own place in the nature of things, then we know that this is believed and proclaimed by Pope John Paul II.

And if Pope John XXIII tells the whole Church with his full papal authority that **"contrary truths cannot exist"** (*Ad Petri Cathedram*, June 29, 1959), then we all believe this to be universally true for all places, for all times and in all situations. A truth and its opposite cannot be true at the same time and in the same sense.

Thus, if known historical facts about the Galileo case are completely suppressed by the same Archbishop who suppressed the researches of Fr. Jean Carmignac, in order to create a totally false 'solution' to the Galileo case, then *no Pope* can be deceived by that or make this deception his own in order to deceive others. Only the Holy Spirit, the "*Spirit of Truth*" (Jo. 16:13), has exclusive use of the Papacy.

All this must be kept in mind when one goes attentively through the findings of the papal Galileo commission, the report of which was handed by Cardinal Paul Poupard to His Holiness Pope John Paul II, who made its contents known to the whole Church on October 31, 1992.

Part Three

A List of Observations Regarding the Address by His Holiness Pope John Paul II On "The Galileo Case"

This papal allocution addresses two distinct topics.

1. *The immediate topic:* the just-concluded weeklong deliberations of the papal Academy of Sciences regarding the problem of *complexity,* (lines 1 - 21; 97 - 117), receiving the usual friendly, fatherly remarks and exhortations, and
2. a *second, remote topic:* the report containing the findings and final conclusions of the papal commission on 'the Galileo case', of which the composition had been overseen, and the conclusions handed to the Pope, by a man who had a proven track record of (to say the least) duplicity (lines 22-96).

Since what this Pope is saying here and now on science in the first and final part of his speech, especially when this is brought into relation with what the Papacy has pronounced on this topic all along, is of *paramount importance* for the true understanding of what is being said in the middle part, we will begin this commentary with *line 9,* section I, subsection B.

9 Here the Holy Father touches on the question of how to reconcile all the individual parts of a complex *'whole'* transcending the totality of all its individual parts.

As has been taught in the past by the Papacy, and will be stressed again here: *science,* by the very *nature* of its being and function, is unable to allocate to itself its place and role in God's creation, the *'whole'* of things. The view and understanding of the 'whole' comes from outside science, and the greater the complexity discovered by science, the more it is in need of 'that greater something outside itself' to view and understand correctly the significance of each individual part. Science may indeed be capable of finding better ways for 'abortion', but it should first know *from outside itself* WHAT IT IS that is being aborted …

For further enlightenment one is referred to the following words spoken by one of Pope John Paul II's predecessors, His Holiness Pope Pius XII:

"As for the most essential problems of scientific knowledge of those whose amplitude embraces its entire realm, those minds which perceive them are, it seems to Us, relatively few in number, and We rejoice at the thought that you are among them. Has not science arrived at the point of demanding that our vision should penetrate readily the most profound realities, and rise to a complete and harmonious view of these **in their wholeness?**"

(Address to the members of the Pontifical Academy of Science., *Modern Science Needs Philosophy*, April 24, 1955).

Here we see Pope Pius XII use exactly the same word as is being used by Pope John Paul II in *line 9* and in exactly the same context.

10 "… to *meta-scientific concepts*, the use of which is, as it were, demanded by the *logic* of his procedure."

The Pope clearly announces here that it is *illogical* to expect science to be capable of explaining the deep link between its own immediate object: the data of the material ('visible') world, and their invisible reality: their *wholeness.* He further states that this illogical character can be appreciated by scientists because it derives from, and is inherent in, their very procedure, the 'scientific method', which demands that, for the explanation of non-scientific matters (i.e. non-observable data), the scientist must go outside science, and so have access to *meta -* (i.e. *'beyond'*) scientific concepts. Finally he strongly implies that it is *illogical* for the Church to surrender Her dogmas to unfinished science, let alone to pseudo-science. (See lines **38, 39** and **40**.)

11 Here the Holy Father states that "*it is useful to state exactly the nature of these* (meta-scientific) *concepts.*" However, he refuses to go into the matter right here other than by declaring in *line 12* that they belong properly to the domain of *philosophy,* after which he proceeds to teach by way of example from *line 13* to *line 16,* concluding once again by highlighting the need for philosophy in *line 17*. This sustained reference to the need of Philosophy, we may add, is the constant theme of the address by Pope Pius XII of April

24, 1955, from which the earlier quoted words as well as the following ones have been taken:

"The triumphs of science are themselves at the origin of the two requirements to which We alluded above.

a) The first task is to penetrate the intimate structure of material beings and to consider the problems connected with the *substantial* foundation of their being and of their action. The question then arises:

'*Can experimental science solve these problems by itself? Do they belong to its domain, and do they come within the field where its research methods can be applied?*'

<u>One must answer in the negative.</u>

It is the *method* of science to take as its starting-point sensations which are external by their very nature, and through them, by means of the process of intelligence, it descends ever more deeply into the hidden recesses of things; but it must halt at a certain point, when questions arise which cannot be settled by means of sense observation.

When the scientist is interpreting experimental data and applying himself to explain phenomena that belong to material nature as such, he needs a light which proceeds in the inverse direction, from the absolute to the relative, from the necessary to the contingent, and which is capable of revealing to him this truth which science is unable to attain by its own methods since it entirely escapes the

Part Three: A List of Observations on The Galileo Case 35

senses. This light is philosophy, that is, the science of general laws which apply to all beings and therefore hold too for the domain of natural sciences, above and beyond the laws discerned empirically.

b) The second requirement springs from the very nature of the human soul, which desires to have a coherent and unified view of truth. If one is satisfied with a juxtaposition of the various subjects of study and their ramifications, as in a kind of mosaic, one gets an anatomical composition of knowledge from which life seems to have departed. Man demands that a breath of living unity enliven the knowledge acquired. It is in this way that science becomes fruitful and culture begets an organic doctrine.

This raises a second question:
'Can science, with the sole means characteristic of it, effect this universal synthesis of thought? And in any case, since knowledge is split up into innumerable sectors, which one, out of so many sciences, is the one capable of realising this synthesis?'

Here again we believe that the nature of science will not allow it to accomplish so universal a synthesis. This synthesis requires a solid and very deep foundation from which it derives its unity and which serves as a basis for the most general truths. The various parts of the edifice thus unified must find in that foundation the elements that make up their essence. A superior force is required for this: unifying by its universality, clear in its depth, solid by its character of absoluteness, efficacious by its necessity. Once again that force is *philosophy*." [End of quote]

We may be allowed to point out here, in support of both Popes, that, what belongs to the invisible (i.e. non-observable) substance, or nature, or essence, of things constitutes their innermost reality, and is 'more real' than their physical attributes, although it lies totally beyond sense perceptions. And so this innermost reality: this substance, nature or essence of all created things is impervious to science (is 'meta-scientific') but not beyond human knowledge and understanding. And thus it is the proper object of another discipline: *philosophy*.

And both Holy Fathers, Pius XII in the above-quoted words and John Paul II in lines 18-21 link the awareness and the appreciation of the true nature of things, that which is 'beyond science', directly to the essential ingredient of culture. For culture to be true, it must be under the direct influence of the Ten Commandments, the Natural Law, true nature of intelligent creation. And once the sharers of that culture are under that benign and beneficial control, they must lift their culture even further by striving constantly for the higher truths. Science and technology cannot bring true culture because they are blind for the true natures of things and so altogether incapable of understanding the true nature of human beings, the "living being", "man", of line 15.

As said before, for the evaluation of what comes next, it is essential that the reader keeps this constant papal teaching on the impotence of science uppermost in mind in an age which has saturated us with the blasphemous idea of the 'omnipotence' of 'god-science'.

After this introduction in his historic address to the members of the Pontifical Academy of Sciences, Pope John Paul II launches

Part Three: A List of Observations on The Galileo Case 37

into the main part: Poupard's report of the 'Galileo case', <u>beginning with *line 22*</u>.

28 *In the future it will be impossible to ignore the Commission's conclusions.*

This calls for the following remark:

This remains equally true in the case of the Commission having deliberately chosen to ignore facts and evidence, as well as in cases that its findings and conclusions are slanted, biased, superficial even fraudulent or in any other way defective or deficient by wilful omissions or faulty interpretations.

That we have reason to suspect *a posteriori* that some of the above-mentioned defects have actually taken place in the composition of the commission's final report, in particular the withholding of known historical facts, and even their wilful suppression, will come out as our commentary goes along.

As for an *a priori* foundation of our suspicions, the reader is referred back to what was discussed at the beginning of this dossier in Part One under the heading: "Why I Mistrust Card. Paul Poupard".

For a good understanding of what comes next, I will print lines 32 and 33 in full.

32 *However, the underlying problems of this (i.e. the Galileo) case concern both the nature of science and the message of faith.*

33 *It is therefore not to be excluded that one day we shall find ourselves in a similar situation, one which will require both sides to have an informed awareness of the field and of the limits of their own competencies.*

It must now become obvious why in the previous pages so much was made of sustained papal teaching on science. *There* the Holy Father was **not** dealing with Paul Poupard's report. But *here*, in dealing with the Galileo case, we have entered its mine-field. Is the 'report' dealing with 'science' in an identical way? ...

And we must bear in mind that a Pope can never be deceived in matters of Faith nor can he ever be a deceiver.

It is to be noted first of all that in *line 32* the juxtaposition is *not* made between science and philosophy, or even between science and theology, but between science **and the message of faith,** the direct domain of papal authority. Consequently, in *line 33*, according to what is being stated here, (*'their'* referring to *line 32)* science is capable not only of imposing limits on philosophy or theology, *'but on the message of Faith'.* That is, science can force Catholic Faith to search for its own limitations. Of this St. Paul has declared:

> "*For Christ did not send me to baptise, but to preach the Good News* (the Message of Faith) *and not to preach that in the terms of philosophy in which the Crucifixion of Christ cannot be expressed.*" (1 Cor. 1:17)

If thus it is revealed truth that the message of Faith cannot be expressed in terms even of philosophy, then it certainly cannot be expressed in terms of 'science' which is subject to philosophy to even understand itself. Philosophy imposes limits on science! From this it follows immediately that science can never be in a position to dictate conditions to, let alone impose any limits on, Catholic Faith.

Furthermore, we know what has been declared by the Magisterium, the highest teaching authority in the Church:

"Of defined doctrine of the Church", (i.e. of the 'message of Catholic Faith'):

a) "*the same <u>sense</u> as used to define it*" (Pope Pius IX in *Inter Gravissimas*, October 28, 1870, quoted by Pope Pius XII in *Humani Generis*, August 12, 1950);
b) "*the same <u>meaning</u>* (Vatican Council I in *De Fide Catholica*, Ch. 4); and
c) "*the same <u>rule of language</u>*", (Pope Paul VI in *Mysterium Fidei*, 1965), "is to be forever retained". To which Pope Paul VI added: "*And let no one presume to change it … under the pretext of new science!*"

To this must be added what Pope John Paul II himself has stated at the beginning of his address in *line 10* <u>when he was not yet dealing with the Poupard report</u>, that, even for the rigorous description and formalisation of the data of experience, the logic of the scientist's procedure demands that he has access to meta-scientific concepts. From which we <u>then</u> drew the conclusion that it must be illogical to expect science to be capable of explaining even its own place in the nature of things. And from which the same logic <u>now</u> compels us to conclude with even greater force that this would render science altogether incapable of imposing limits on supernatural things. "<u>Let no one presume to change it</u> … <u>under the pretext of a new science!</u>"

Taking all these observations together, having especially in mind the sustained papal teaching on the limitations of science, and even on its impotence, it is obvious, therefore, that the Holy Father is reading here from a prepared text written by a Teilhardian with the future rehabilitation of Teilhard de Chardin uppermost in mind. And in *lines 5, 25, 72 and 75,* the Holy Spirit made His chosen instrument Pope John Paul II unmask the author of this text by making him point the finger directly to the man he had put in charge of the final coordination of this study commission, Cardinal Paul Poupard, Archbishop of Paris. But this appointment, whenever it may have taken place, could *never* be divorced from the historical events of 1981, the very year when the trust of the Holy Father had been betrayed on the day of the shooting, May 13. For it was on that day that Cardinal Paul Poupard became the recipient of a laudatory telegram sent by the Secretary of State, Cardinal Casaroli, congratulating him <u>with the forthcoming centenary celebrations of the birth of the late Teilhard de Chardin</u>, which had been strictly forbidden by Pope John Paul II barely two months earlier ...

If, from the subsequent corrective action he took, the Holy Father showed the whole Church that he was fully aware of the duplicity that had taken place, why then did he see fit to appoint Cardinal Paul Poupard at the head of this sensitive commission for the final coordination of, and the subsequent reporting on, its findings?

Let us thank the Holy Spirit that the Pope had the wisdom and the foresight to do so by considering the following.

If a man above reproach had been appointed, and another betrayal of trust would subsequently have taken place, it would maybe not have been all that hard to uncover, but would have been exceedingly hard to pinpoint and to bring home to a particular individual.

If however an untrustworthy teilhardian is placed in this particular position of trust, there are, in theory and in reality, two possible outcomes.

(i) either the man in charge has had a genuine change of heart which would come to light in the utter quality of the final submissions to the Holy See;

(ii) or else he sees in his appointment a golden opportunity to further the cause of teilhardism within the Catholic Church, which would result in a mongrel of a report being presented to the Holy Father.

By reading the speech substantially as it had been prepared for him by the head of the Commission in line with its findings, the Holy Father has truly solved the problem for us by letting the whole Church know that a third betrayal of trust has indeed taken place. And the whole Church is now placed in the best possible position to ascertain what we said in our remark after *line 28*: how well do the findings of the commission stand up against the inexorable Truth of Tradition and of any historical facts?

Bearing in mind then what we have said before: that the exclusive use of the Papacy is the dissemination of truth as seen in the light of the Holy Spirit, let us now go and find out how the Holy

Spirit fouled up Cardinal Poupard's commission, by examining how well, what this commission presented to the Holy Father, *as reflected in his speech*, squares off with the known historical facts of "the Galileo Case". This we do on the force by which Pope Pius XII wrote in *Humani Generis* (1950):

"*He* (the Christian) *will weigh it* (the latest fantasy) *carefully... making sure **that he does not lose hold of the truth already in his possession or contaminate it in any way ...**"

What occupies us here are the answers to the following two questions.

(i) Was a *true science* presented to the Holy Church by Galileo, and
(ii) Did <u>all</u> the *known facts* (the *truths* spoken of by Pope Pius XII) surrounding the 'Galileo case' find their way into the final report presented to the Holy Father by Card. Paul Poupard?

Answer to (i).

NO. <u>Not</u> according to the present Holy Father, Pope John Paul II. Here the reader is once again referred to what the Holy Father said in lines **38, 39 and 40** of his address to the PAS.

And again, **NO.** <u>Not</u> according to the historical truths that lie irrefutably exposed in the pages of history. For a good understanding of this it is necessary to review in some detail:

Part Three: A List of Observations on The Galileo Case 43

The Copernican System

Books have been written about it. Of the welter of information contained therein, Poupard has refused to submit to the Holy Father the most salient historical facts which alone, in their accuracy, are capable of shedding the most brilliant light on "the Galileo case" as it was seen and appreciated by the Holy See and seen through the far-seeing eyes of God's chosen instrument at that time, *St. Robert Bellarmine.*

1. The one whose name has been given to it, *Nicholas Koppernigk,* a Canon of the chapter of Frauenburg cathedral on the Baltic coast, and author of *The book of the Revolutions'* refused all his life to publish it knowing it to be unsound. It was finally published by his friends, and the first copy was presented to him the day he died of cerebral haemorrhage. Laid on his death bed, all he could do was stroke it ...

2. *The Copernican System* is a purely abstract mathematical model composed of highly imaginary epicycles for the paths of some of the planets and, needless to say, totally devoid of any semblance of reality.

   Some 30 years earlier the Canon had allowed an abridged version of his book to appear, but only handwritten and sent to a highly selective group of scientists. Its impact was dismal, but it became the cause of rumour and hearsay, the stuff that the enemies of the Church forge their weapons from.

This abridged version contains Copernicus' 7 Revolutionary Axioms, with 7 short descriptions without proof or mathematical demonstrations. The few selective readers who had access to it, were referred "to his forthcoming book" for the latter. They had to wait for 30 years. The last paragraph of the treatise proudly announces:

"*Thus Mercury runs on 7 circles in all; Venus on 5; the Earth on 3, and around it the Moon on 4; and finally Mars, Jupiter and Saturn on 5 each. Altogether, therefore, 34 circles suffice to explain the entire structure of the Universe and the entire ballet of the planets.*"

It is absolutely essential for the understanding of the future controversy that good notice is taken of this tally of '<u>altogether 34 circles suffice</u>'. For the 'abridged version' appears to have merely been an optimistic preliminary announcement. When the good Canon gets down to detail in his *Revolutions*, he was forced to add more and more wheels to his machinery <u>and the number of epicycles grew to 48</u>! And nowhere in his book does Copernicus either give an explanation for the discrepancy or a retraction of the original miscalculations. Even men of science show unwittingly that they never read *The Book of Revolutions* if they only adhere to '34 circles' …

Needless to emphasise already here right at the beginning when there was absolutely no animosity around, that no self-respecting Church could ever contemplate to allow

Dogma and Exegesis to give way before the claims of such a scientific non-starter.

3. *The Copernican system has no physical centre ...*

How maddeningly frustrating the *Book of Revolutions* is, comes clearly to light from the following:

Copernicus' book begins with a broad outline of the theory. This synopsis occupies less than 20 pages, or about 5% of the whole. And the remaining 95% consists of the application of the theory. And when that is completed there is hardly anything left of the original doctrine: it has, so to speak, destroyed itself in the process ... At the beginning of the book Copernicus had stated:

"...in the midst of all dwells the sun, the centre of the universe. Sitting on the royal throne he rules the family of planets which turn around him .... We find in this arrangement an admirable harmony of the world."

But in Book III, when it comes to reconciling the doctrine with actual observation, *the earth no longer turns around the sun,* but round a point in space removed from the sun by a distance of about 3 times the sun's diameter. Nor do the planets revolve around the sun - as every schoolboy believes that Copernicus taught. The planets move on epicycles of epicycles, centred not on the sun, but on the centre of the EARTH's orbit. Thus there are **two** 'royal thrones': the sun, and that imaginary point in space around which the earth moves. The year, that is the duration of the earth's complete revolution round the Sun, has a decisive influence on the motions of all other planets. In short, the earth appears equal in importance in governing the solar system to the sun itself.

No wonder the poor Canon ended up pleasing no one: neither Aristotelian cosmology, nor the Ptolemaic System, nor the reconciliation of these two, nor, as is infinitely more important, KEPLER.

4. *'No ecclesiastical animosity'*

Now, after 450 years of controversy surrounding the Galileo case, it is almost as impossible to visualise the total absence of any ecclesiastical animosity against the man who held and discreetly diffused his researches, as it is to accept that there was actual ecclesiastical encouragement, even insistence, to publish them! For the Churchmen knew as well as Copernicus himself, that the scholarly Canon was neither the first man on earth to hold such ideas nor the only one who was engaged in their research at that time. If 'the Copernican system' spread by osmosis, this was not the wish of the Church. For the Church has nothing to fear from genuine scientific research ...

5. *Galileo never read Copernicus' book ...*

... at least not before his two 'trials'. All his life this mathematician from Padua shared the same fear as the timid Canon: the fear of ridicule, "of being hissed off the stage" as he puts it himself. Neither did Galileo ever bother to read Kepler's works containing the latter's discoveries of the three laws of planetary motion.

6. *Galileo was no astronomer.*

This is revealed by the first recorded contact between Galileo and Kepler, a - for the 'Galileo case - most important letter Galileo wrote to thank Kepler for his book, *Cosmic Mystery*. The letter, dated August 4, 1597, is highly revealing for several reasons. But, when compared with a 1606 manuscript of a treatise written by Galileo for private circulation amongst pupils and friends, the letter is no longer only 'highly revealing', but becomes positively shattering … (The reader is referred to Appendix A for part of this letter). The letter is important for several reasons.

Firstly, by showing the fact that Galileo accepted Kepler's first book, an astronomical hoax, at face value, the letter strengthens the conviction of many moderns, *that Galileo was not a real astronomer at all* but simply one who, amongst other things, trained his 'spy-glass' on the sky and discovered various curios. He never left us a 'system', never did any calculations, held firmly to Copernicus' impossible system in private, taught its opposite in public, and even his sparse star maps are grossly inaccurate. His undoubted talents lay altogether elsewhere.

7. **Galileo was highly contradictory.**

"For many years" he was a self-professed convert to the Copernican system in strict privacy, but a public refuter of the system in his university lectures.

<u>Secondly</u>, the letter to Kepler provides us with conclusive evidence that Galileo had become a convinced Copernican in his early years without ever having read the Canon's book ... He was 33 when he wrote the letter to Kepler, and the phrase *"many years ago"* indicates that his 'conversion' took place in his twenties. Yet his first explicit public pronouncement in favour of the Copernican system was only made in 1613, 16 years after his letter to Kepler, when Galileo was 49 years of age. Through all these years he not only taught, in his public lectures, the old astronomy according to Ptolemy, but *expressly repudiated Copernicus!* In a treatise which he wrote for circulation among pupils and friends, of which a manuscript copy, dated 1606, survives, he *adduced* all the traditional arguments *against* the earth's motion, arguments which, if the *1597* letter to Kepler is to be believed, he himself had refuted "many years before".

But the letter is also interesting for other reasons. In a single breath, Galileo four times evokes Truth, then calmly announces his intention and his practice of suppressing Truth.

Why, in contrast to Kepler, but in line with Canon Koppernigk, was he so afraid of publishing his opinions? He had, at that time, no more reason to fear religious persecution than Copernicus had. In Copernicus' own day, Catholics were favourably inclined towards him, going by the fact that Cardinal Schoenberg and Bishop Giese had urged him to publish his book. Twenty years after its publication, the Council of Trent redefined Church doctrine, but it had nothing to say against the heliocentric system of the universe. Galileo himself enjoyed the active support of a galaxy of Cardinals, including the future Urban VIII, and of the leading astronomers among the Jesuits. Up to the fateful year 1616, discussion of the Copernican system was not only permitted, but encouraged by them. Under the one proviso, that it should be confined to the language of science.

This Galileo refused to do …

8. **The Copernican model is <u>not</u> a heliocentric system.**

Apart from the impossibility of "epicycles on epicycles" as physical quantities with regard to planetary motion: contrary to the most persistent popular (and as it turns out, also academic) belief, both now as in the time of the Renaissance, the Copernican system is *not* a truly heliocentric system as Canon Koppernigk distinctly teaches that the sun ('*helios*' in Greek) is **not** the centre of the planets' orbits. Neither is this true for their elliptical paths in the calculations of Kepler.

9. Galileo's downfall

Although he had access to Kepler's rigorous mathematical calculations since the year of their publication, 1609, Galileo ignored them and began to insist in earnest around 1616, and persisted until after 1633, that the Holy Church should modify Her theology and exegesis precisely, *and* exclusively on the strength of this impossible mathematical and scientific non-starter, *the Copernican system,* without any need for proof. Luckily for him, Galileo not only had the greatest Catholic mind of the day <u>as opponent</u> to these preposterous extortions but also one of the greatest Saints of the Renaissance <u>as friend</u> to guide him back to sanity and to the road to Heaven
Answer to (ii) above.

None of these historically verifiable data come out in the address of the Holy Father on October 31, 1992. So we must assume that they have been deliberately suppressed from the *'ten year* findings' of the papal commission and were consequently omitted from the 'report' handed to the Holy Father, from which the papal speech was subsequently prepared ...

Only Teilhardians need the corrupted version of the 'Galileo case' for the 'restoration' of their fallen idol, Teilhard de Chardin. But since it is public knowledge that

1) that the Holy Father, two months before the May 13 shooting in St. Peter's Square, expressly forbade the

Part Three: A List of Observations on The Galileo Case   51

sending of any laudatory telegram from the Vatican to Card. Poupard on the occasion of his preparations for the centenary celebrations of Teilhard de Chardin's birth, and

2) that the Holy Father personally ordered the public correction in the *Osservatore Romano* of the laudatory telegram that was sent in his name on the day of the shooting by his secretary of State, Card. Casaroli, to the Archbishop of Paris, Card. Paul Poupard, by the forced inclusion of Pope John XXIII's 1962 *Monita* against Teilhard de Chardin,

we have enough information to know for certain that the Holy Father is *not* a Teilhardian, and that he is *not* in favour of Paul Poupard's strategy of having a slanted report serve the rehabilitation of one Teilhard de Chardin ...

We continue to search out what Card. Poupard's commission submitted to the Holy Father as revealed by the latter's speech.

41 It is here where the report starts to weave its web of confusing historical facts with make-believe, and where it begins to drive the thin end of the wedge into the whole argument, in order to dislodge the truth from its absolute throne (*line* 56) by separating historical facts from the knowledge of them, and by misdirecting the thrust they possess away from "*the wholesome streams of Thomism*" with its so necessary distinctions (Pope Leo XIII) into the everlasting desert of Existentialism.

But it is also here that we pick up the deadly weapon given to us by Thomism and the Papacy: *theory* is open to reinterpretation in the light of evidence and known facts. Knowing the full force and weight of this, Paul Poupard's commission has been at pains to distort and suppress known historical facts and evidence in relation to the Galileo case just as, during that same time span, the 1980s, his Institut Catholique on his orders had suppressed the mounting evidence of Fr. Carmignac's research.

42 After the report had turned the whole argument upside down, and before this is realised, it is now necessary to speedily slip in the conclusion essential to the central thrust of the modern counterpart of Galileo, our Teilhardians, to create the impression 'that it was the theologians who had a problem'...

This 'message' must be weighed against overwhelming facts to the contrary.

The Church of all times has no problem whatsoever either with true science or with pseudo-science. If She listens attentively to true science, She makes short shrift with scientism.

Her greatest champion in Galileo's days, St. Robert Bellarmine, had no problem with Galileo's pseudo-science, the Copernican system. He simply refused to accept this unproven attempt as an issue fit for demanding a change in Catholic teaching. Why? Let us go to a few admissions in Pope John Paul's address.

38 "Galileo made no distinction ..."

39 "That is why he rejected the suggestion made to him to present the Copernican system as a hypothesis, inasmuch as it had *not* been confirmed by irrefutable proof."

40 "Such was an exigency ('urgent or pressing requirement') of the experimental method of which he was the inspired founder."

Part Three: A List of Observations on The Galileo Case 53

From this sober resumé by the Pope, and from the detailed historical facts at anyone's disposal in relation to line 39, which of the two protagonists stands before the Bar of History with 'the problem'?

Galileo stands before the Bar of History with a 'system' that is not only in reality, but even mathematically untenable, physically absurd and astronomically impossible.

He stands before the Bar of History as a living contradiction: a man who claims that privately he was a convinced Copernican for years while teaching its opposite in public.

He stands before the Bar of History as being *no* astronomer.

He stands before the Bar of History as unwilling to make the necessary and correct distinctions, as refusing to present an impossible system merely as a working hypothesis against the urgent requirements of the experimental method he is supposed to have 'invented'.

Against all this welter of scientific and historical evidence, Paul Poupard's commission dares to assert as its 'findings': "that it was the theologians who had the problem".

Let us be grateful to Almighty God that, on the claims and insistence of the historical figure who gave his name to this case, 'the Galileo case', that central to that case on his own insistence, is a mathematical chimera, a physical absurdity and an astronomical impossibility: the Copernican system …

43 "Thus the new science obliged theologians …"

After all the obstacles have been removed in the *ten years* it took to fabricate the report, the final slip-in now is to baptise the pseudo-science of 'the Copernican system' and to call it loosely by its new name: "*the new science* ..." This is necessary to make out that, after 350 years, and after the suppression of historical facts that could be held up against this, "the new science" has acquired enough status in the intervening period of time to claim what St. Robert Bellarmine would not allow it to claim: the *right* to impose limits on the message of Faith ...

St. Bellarmine was correct when he reassured theologians not to feel obliged to take a *non-science* as a serious threat to Catholic theology, exegesis and Faith.

44 "Most of them did not know how to do this."

That is why the first 'trial' of Galileo was of such great importance: St. Bellarmine made the Church's principle crystal clear that no theologian needed to feel ashamed or guilty if he was unable to answer the preposterous demands and claims made by Galileo, since they came from a totally unproven 'science'. If the Church was prepared to wait until proof was forthcoming, why should ordinary theologians be blamed if they followed that wise lead and refused to be drawn into futile arguments which centred mainly around the hypothetical postulates of a pseudo-science? Exactly as is now the case with 'evolution'. But just as in the early 17th century Galileo Galilei would leave no stone unturned to have the Copernican system accepted by the Church as a genuine fact of science without the cumbersome necessity of providing proof, so now will the Teilhardians and Modernists resort to anything, even to the falsification of reports to the Holy Father, to see evolution accepted as "*the new science*" in line with *line 43*.

And so we could go on. The central message of this 'report', as reflected in the papal speech, appears to be that Galileo was right

Part Three: A List of Observations on The Galileo Case 55

and the theologians were wrong, that Galileo "showed himself more perceptive in this regard than the theologians" (line 45). Nowhere are the theologians given the *right*, not even with the exclusion of St. Robert Bellarmine, to distance themselves from the debate because of its futility inherent in its false premises. This becomes especially clear in line 49.

49 Poupard and his 'commission' are obviously pitching here for Teilhard de Chardin's theistic evolution, a mathematical, physical, biological, philosophical, theological, cultural and religious absurdity: the classical non-starter. So, moving away from science, he wants this chimera presented and accepted simply as *"the birth of a new way of approaching the study of natural phenomena"*. No matter how unproven and even preposterous the claims of such a "new movement" may be, all that is needed is that somehow it appears as "a new way of approaching the study of natural phenomena" and it automatically acquires the same status as science without the cumbersome need of having to prove its scientific worth.

And now comes the rub. According to Poupard and what his 'commission' uncovered from a totally biased and slanted Galileo case, such a new-comer acquires automatically an inherent *right* <u>to demand that the Church clarifies Her position towards it</u>, or leaves the whole field by default to "this new movement", exactly as Galileo had put this to St. Bellarmine, for which the Saint gently but ever so sternly rebuked him. Meanwhile, the language of the commission's report to the Holy See, as reflected in the papal speech, becomes more demanding:

50 "It <u>obliges</u> them …"

51 " … this new way <u>requires</u> …"

52 "The upset caused by the Copernican system (a mathematical hybrid and a scientific non-starter) thus <u>demanded</u> (sic!) epistemological reflection on biblical sciences": something St. Robert Bellarmine refused point blanc to hold the Holy Catholic Church bounded to.

57 " ... a new scientific <u>datum</u> ... "

Neither 'the Copernican revolution' nor 'evolution' can, not even with the wildest stretch of the imagination, be called a *datum*, a proven scientific *fact*. And so pastoral prudence requires that their inherent contradiction of the truths of faith be sternly dismissed <u>as not coming from true science</u>.

After I had reprinted the speech line by line, it may strike the thoughtful reader as it struck me that, first of all, 'the Galileo case' was presented extremely slanted and biased. That put me immediately on guard: here was a deliberate effort being made to tie the Holy Father to a particular viewpoint, and a false one at that. This led very quickly to two conclusions:

1. The Holy Father was virtually reading a report.
2. And remembering that it had been Cardinal Paul Poupard who had betrayed the Holy Father in 1981, the very year of his appointment to the commission's top position, it occurred to me that he was at it again. His appointment had been a deliberate test by the Holy Father: to expect from Paul Poupard a genuine conversion. Or not? From then on things started to fall into place. I became convinced that the hoped-for 'conversion' had not taken place; that the commission had not been genuine but had come together with

'ulterior motives'; that therefore the Holy Father was not teaching the contents of this part of the report: he was simply reading it out aloud, and thus that the specific teaching must have been something else, namely to reveal to the whole Church that <u>a third Poupard betrayal</u> was in the process of taking place.

But there is more to the commission's document than bias, suppression and slanted reporting. According to what the Holy Father read out, it contains contradictions. In lines *38, 39* and *40*, the Pope clearly announces, in line with St. Robert Bellarmine, that Galileo had **not** produced '*a new science*'. What he brought up lacked all the criteria of a rigorous, irrefutable and proven reality, and could at most only be classed as a theory, which, again according to the Pope, *Galileo refused to do*. Yet all through the report it is referred to as '*the new science*', not only demanding that the Church made Her attitude to this new science known, but also demanding that She accommodates to it the teaching of *the message of Faith* (lines *32* and *33*).

Studying the 'report' from this angle soon revealed the 'ulterior motives' of the commission: *there was a hell of a lot more in it than just Galileo*. It was crammed with all the things a Teilhardian like Poupard would love to hear the Pope proclaim. These I have enumerated. To Poupard's commission, the 'Galileo case' is nothing but a lever to achieve another 'rehabilitation'.

Thus I was slowly led to the full glory of our Catholic Faith in the Papacy as expressed above, and what it is exactly that the Holy Spirit reveals by such a speech and by such attempts at subversion.

And how tables are being turned on teilhardian conspirators through the line-by-line fine tuning of the exact steps by which apostasy and transition into the One-World 'church' are being envisaged by them. All we have to do is study those areas carefully, now that we have been officially warned by the Holy Spirit and His infallible instrument.

Thus the important things the Teilhardians wanted to be read out by a Pope *were indeed read* out, but for a totally different purpose than they had originally intended. They were not taught: they were *exposed* with the infallible authority of the Holy See by the Only One Who has exclusive use over that See and its genuine occupant. We are now in the possession of a *blueprint of subversion,* drawn up by Teilhardians, but, at their own behest yet to their everlasting downfall, they were held up to the Church by none other than the Holy Father.

Only a Pope could read this out to us. No one below him would have the authority. And then it would no longer matter if he told each one of us individually in name of the Holy Catholic Church, or else if he told the whole Church in name of the Holy Spirit. Either the Holy Church had known all along and now has made it known to us through Her visible head, or else the Holy Spirit alone knew, and has now made it known to the whole Church through the Pope.

In such a scenario the whole question of *true science* and *true exegesis* is saved, that is, they are being preserved from becoming vehicles for this transition into apostasy.

To highlight to the reader how tremendously important to Teilhardians the Galileo case is in their all-out drive to see any 'science' accepted as *the number one power* to limit *the message of faith*, I conclude this dossier with two quotes from Archbishop R. Weakland.

The first one of these appeared in the *New Yorker* of July 15, 1992:

"The major issue facing the Church is Sexuality: that is 'the big one'. We still have not come to terms with it all. Our *scientific understanding* is still rudimentary. Normal sex drives, homosexuality, paedophilia, contraception - the fact that science can control or modify human reproduction. It is a great and frightening frontier, the *Galileo case of our day*. And the Church is reluctant to accept *the results of the human sciences:* instead it harks back to the days when you could say, 'This is black, this is white; this right, this wrong'."

That is the <u>American</u> voice of betrayal …

But here we have it echoed, undiluted and in a nutshell, what Card. Poupard made the Holy Father read out for the *advancement* of … what? If this must be called 'betrayal' here, then Poupard's effort must also be branded as the <u>European</u> voice for the *advancement of betrayal.* However, by the direct intervention of the Holy Spirit, this became instead the exact opposite: the Holy Church making her visible head read it all out, word for word, for the *exposure* of betrayal …

If that was the American voice of betrayal *before the* Holy Father gave his Galileo speech in 1992, then how does that voice now sound?

Here it is, taken from the editorial in *The New York Times* of December 6, 1992, which may truly be called his 'manifesto of apostasy'. This reprint is from *The Wanderer*, December 17, 1992.

"… Weakland also referred to the Holy See's recent pronouncements on the Galileo case as 'a symbol of such a new peace treaty' with the modern world, and asserted that women's ordination could be its *'new Galileo'*, a situation that could be avoided if the Church accepts *'the new insights of anthropology, psychology, and sociology'* embraces *'the startling discoveries of science'* and takes a *'growing leadership role - equally through its female and male believers in the new global culture'*. Such a project must be ecumenical and interfaith."

So there it is. The final 'dénouement' of the whole teilhardian revolt: the Church being 'reminded' of this new teaching, and the Holy Father being quoted as having recently taught this new teaching, that everything: Faith, the Sacraments, teaching, Church and culture are ultimately all controlled by 'science'. Just as Poupard 'discovered', according to his final 'report', <u>that it was only after the Church had accepted this 'fact-of-life', that finally the dust had been allowed to settle on the first Galileo case</u> …

"*You shall crawl on your belly **and** eat dust every day of your life*" (Gen. 3:14).

If the Holy Father is bound to follow in his Master's footsteps all along the *Via Dolorosa* of Our Lord's Passion, then he will look weak but he will again be stronger than any other man on earth. Anyone who will allow himself to be deceived by this appearance of 'weakness' will desert him. Anyone who will allow himself to be led into thinking that this 'weakness' must now be taken as a reversal of his former strength will be scandalised by the Holy Father, and because of that, will part company with him in his incomprehension.

If now the Holy Father is following in Christ's footsteps, then this can only be because it is the Church's wish to suffer that same fate, crown of thorns and all. He is the visible and accurate barometer of the Church's mystical passion. But we should not only watch this barometer: we should be part of its graduations. Our Lord's Passion was surrounded by incomprehension and by pure mischief-making. So will the Church's, the Holy Father's, and ours

…

Appendix A

Part of Galileo's letter to Kepler, August 4, 1597

"... and indeed I congratulate myself on having an associate in the study of Truth who is a friend of the Truth ... but to congratulate you on the ingenious arguments you found in proof of the Truth ... I promise to read your book in tranquillity, certain to find the most admirable things in it, and this I shall do the more gladly, as I adopted the teaching of Copernicus many years ago, and this point of view enables me to explain many phenomena in nature ... I have written many arguments in support of him and in refuting the opposite view, which however I have not dared to bring into the public light, frightened by the fate of our teacher, Copernicus himself, who, though he acquired immortal fame with some, is yet to an infinite multitude of others an object of ridicule and derision."

The reader is reminded that Kepler's book referred to here is *not* his master-piece, *A New Astronomy*, which was published 12 years later in 1609 and contained the first two of his *laws of planetary motion*. (The third law was published in 1619). The book referred to here was an example of Kepler's lifelong 'dream world', *Cosmic Mystery*, in which, at the age of 26, he tried to prove that there can only be 6 planets because there are only 5 Pythagorean solids and, placed in a circle, 5 intervals between them, into each one Kepler puts, with the greatest of labours, a different solid, his scientific hoax.

Appendix B

Excerpts from the Letter of Card. Bellarmine to Fr. Paul Anthony Foscarini, O. Carm.
of April 12, 1615.

My Very Rev. Father,

"… As you ask for my opinion, I will give it …

First. It seems to me that your Reverence and Signor Galileo would act prudently were you to content yourselves by speaking hypothetically and not absolutely, as I have always believed that Copernicus spoke. To say that on the supposition of the earth's movement and the sun's immobility all the celestial appearances are better explained than by the theory of eccentrics and epicycles, is to speak with excellent good sense and to run no risk whatever. Such a manner of speaking is enough for a mathematician …

Second, As you are aware, the Council of Trent forbids the interpretation of the Scriptures in a way contrary to the common opinion of the holy Fathers …

Third. If there were a real proof that the sun is in the centre of the universe, that the earth is in the third sphere and that the sun does not go round the earth, but the earth round the sun, then we should have to proceed with great circumspection in explaining passages of Scripture which appear to teach the contrary, and rather admit that we did not understand them than declare an opinion to be false which is proved to be true. But as for myself, I shall not believe that there are such proofs until they are shown to me.

Nor is a proof that, if the sun be *supposed* at the centre of the universe and the earth in the third sphere, the celestial appearances are thereby explained, equivalent to a proof that the sun actually *is* in the centre and the earth in the third sphere. The first kind of proof might, I believe, be found, but as for the second kind, I have the very gravest doubts, and in the case of doubt we ought not to abandon the interpretation of the sacred text as given by the holy Fathers ..."

(From *Robert Bellarmine, Saint and Scholar,* 1961, by James Broderick, S.J.).

From the quotes which appear in this paper it has become manifest that the sentiments expressed in this letter are those of the papacy in general and of Pope Pius XII in particular.

Epilogue

In 1991 I wrote a little booklet called *The Hedge of the Vine*. The inspiration to write it came from a text found in *Aeterni Patris* an encyclical written by *Pope Leo XIII* in 1879 for the timely restoration of the serious study of the 'Everlasting Philosophy' of St. Thomas Aquinas. It is in this encyclical that this far-sighted Holy Father graces the power and clear-thinking of this pinnacle of human thinking with this glorious name: *"the Hedge of the Vine"*.

[Ed. *The Hedge of the Vine* is published in *The Selected Works of Frits Albers, Volume 3*. En Route Books and Media, 2024.]

The *Vine* is pure and childlike Faith in God, planted by God on earth in Paradise, when He infused this supernatural, precious Gift in our Proto-Parents from the moment of their first and spotless existence. After the First Lapse this hardy yet delicate Vine was preserved in 'the strain of God'; was carried by Noah over the waters of the Flood and beyond, until it received its deepest roots yet in the Faith of Abraham. From then on it blossomed out in the Jewish salvation history until it peaked at its all-time high: the Faith of the Blessed Virgin Mary, the New Eve producing in Her virginal womb the New Adam of the Redemption, the Son of God Himself. From then on it bore fruit in abundance all over the world in the never-ending riches of the Faith of the One, Holy, Catholic and Apostolic Church, created in the image and likeness of the Holy Mother of God.

In this way the 'hedge' becomes *"the love of Truth"* by which the human mind surrounds and shields the Vine, protecting it as by an impenetrable bulwark. Of this 'love of Truth' St. Paul has sounded the dire warning for those who are found without it in the time of Antichrist:

"*... for they did not possess the 'love of Truth' which could have saved them*". (2 Thess. 2:10)

By accepting the frightful lies and deceptions of Satan and of the world at that time *as truth,* they dismantled their hedge and exposed their Vine ... with the dire consequence that another warning of Sacred Scripture finds its fulfilment in their new existence:

"He that uproots a hedge, a snake will bite him". (Eccl.10:8)

And we all know what 'serpent' the Holy Bible is talking about here: the one that bit Eve, when she helped him to uproot the 'hedge' of her clear thinking around the Vine of Her Supernatural Faith in God ...

From this little book I may be allowed to quote the following few lines in defence of this present paper.

[Quote from *The Hedge of the Vine*]

"As these pages have made clear, anything or anyone who wishes to have access to the 'Vine' i.e. the Faith of a Catholic, is first referred to the 'Gate in the Hedge'. There, in the presence of Christ 'Who is both the Gate and the Truth', credentials are checked by the glorious human intellect against truths already ascertained. As Pope Pius XII puts it:

"*... so as not to light on something which contradicts truths already ascertained. The Christian will weigh the latest fantasy care-*

fully, making sure that he does not lose hold of truths already in his possession, or contaminate them in any way with great danger and perhaps great loss to the Faith itself." (Humani Generis, 1950.)

And if anything is being proposed that does contradict any truth, whether ordinary or Revealed, already in a Catholic's possession, it will be barred at the gate.

Now comes the crunch!

In order to get yet into that Catholic's mind what has been rejected at the gate, *violence* will have to be done to the intellect which by nature is attuned only to the truth. And that is how the hedge gets pulled up by the owner with the help of his outside 'friends'; how the Vine gets exposed and trampled underfoot, and how the owner's mind and Faith get bitten by a snake ... or as Pope Pius XII put it: *become contaminated.*

But there is one exception.

There is one person on earth for whom, if he presents himself at the gate, even Our Lord steps back to let him in without the checking of credentials. There is one person on earth who has direct access to the Vine, because he will *never* do violence to it! And if this man by the name of "*Peter the Rock*" firmly declares "*that the use of contraceptives is intrinsically evil*" then he is allowed to come in and graft this information onto the Vine. And instead of him giving reasons to the intellect, the human mind will accept the new belief, will continue its role, and, as the hedge of the Vine, will find reasons to defend the new graft on the Vine. And if this same man presents himself again, and declares "*that the Novus Ordo Missae is the new format of the Holy Sacrifice of the Mass for our times*" then we know that he says that with the authority of the Man at the

Gate, Who let him in because the information carried by 'Peter the Rock' is completely in tune with the Truth the Man at the gate happens to be in Person."

[End of quote from *The Hedge of the Vine*]

And so, on the 31st of October, 1992, the Holy Father presented himself once again at the gate of my hedge with information he knew could not possibly be grafted onto the Vine of Faith in a Catholic's mind, since, as Pope Pius XII had said, it contained things *"contradicting truths already ascertained"* So, at his own invitation, I sat with him there and then, going through his 'Galileo speech' line by line.

Several difficulties against this procedure could be raised here. Some of the usual ones could be:

1. If the Holy Father wanted the contents of his Galileo address carefully weighed *"so as not to contaminate truths already in a Catholic's possession,* why then did he not say openly and plainly what this paper assumes was the thrust of his Galileo address?
2. How can anyone avoid the trap of sitting in private judgement over what the Holy Father says?
3. Where is, with a paper like this one presented here, our safety shelter of final recourse: that we take what the Holy Father says without questioning? "You know what I mean! That saying about when the Holy Father declares 'white' what we may think is 'black' ...".

Epilogue

No. 1 I dismiss as un-thomistic. We can never fruitfully argue about a non-reality, about something that did not happen. We only have as fact the reality of his speech as it was given. Only that can fruitfully be the centre of discussion. We may speculate over this fact, ponder about e.g. *why* the Holy Father did say what he said. But we may not dismiss an ordinary analysis simply because it does not address itself to an alternative that did not take place.

If then someone insists: "OK, I accept that, but then let me reframe the question from the negative to the positive:

"Why then did the Holy Father, according to your paper, say in a roundabout sort of a way what your paper claims he did say?"

Before the obvious answer will be given, let it first be pointed out that going through a papal speech line-by-line is *not* an insult to the Holy Father: *it is honouring him!* It shows that what he has to say is considered important enough to deserve very careful attention and study.

Had he publicly dismissed the findings of the Poupard commission on Galileo as they deserve to be dismissed, there would have followed - in the hostile climate and mistrust caused by rampant Modernism within the confines of the Holy Church today - at least ten years of endless arguments and bitter bickering over which aspects of the findings he did not like. Under the circumstances it would be far better to show the whole Church: "*Listen to* **this**! *This is the package I received from my Galileo commission*". Those who understand him perfectly only need 'half a word'. And all those Modernists and Teilhardians who are too stupid to see that they have been had, cannot blame him for what he did: he did exactly what they expected him to do, <u>he read out what was given to him</u> …

No. 3 is easy to deal with: I *do not question* what Pope John Paul II said in his Galileo speech. Much of what this address contains had already been held up as 'black', i.e. untrue, by well-known facts of history, or by the Church through Her previous Popes or Councils. My paper agrees with this, it brings that out. And so I remain adamant that no Pope is now, or ever was, or ever will be, declaring 'white' what other Popes have declared 'black'. I did not query a single statement: I only attached to it the verdict of history; either well-known historical facts, or well-known historical teaching.

That leaves No. 2: the question of 'sitting in private judgement over what the Holy Father teaches'.

Whenever the Holy Father teaches, he teaches the truth, addressing both the intellect as well as the Catholic Faith of his flock. These truths must he understood, accepted and believed. I will never hold that the Holy Father did not teach in his Galileo address. But if he did *not* teach the lies, deceptions and suppression of known historical facts and of known papal teaching as were contained in the Poupard report, he *must* have taught something else.

This time, when Pope John Paul II presented himself at the gate of *my* hedge, the cargo he brought with him had a definite smell about it. But refusing to do what so many other Catholics may have done in this case: take a cursory look at his Galileo address and then show a hasty disinterest and a speedy forgetfulness in order to avoid an acute mental conflict, I not only let him in, but I did him the honour of sitting with him to go through his Galileo address

Epilogue

line by line, in order to fully understand *which Truth* he was teaching, and exactly *which truth* he wanted me to accept and believe. In the full knowledge that he would never graft onto my Vine as true what had previously been taught as not being true, either by his own mouth or, what is the same, by the mouth of the Papacy and of the Councils of the Holy Church, I was assured that the exercise would be totally free from any acute mental trauma.

Must it now be taken that it would have been better *not* to have looked too closely at the Holy Father's Galileo address so as to avoid an acute mental conflict? <u>The conflict of having to accept that the lies and the suppression of truth held up in this address have somehow acquired the status of being true</u>, because, being *read out* by the Pope to the whole Church, they have taken on the appearance of being somehow *taught* by him? (Weakland). Is it necessary to feel unhappy or uneasy when his Galileo address is studied closely? Could it be that the decision not to look too closely is due to the fact that a private 'tranquilliser' has been found assuring the soul that, what formerly was held up to us as 'lies', has for some mysterious reason somehow ceased to be lies? If (from a cursory acquaintance with this Galileo address) a Catholic acquired a premonition of an impending conflict between the general principle "of always adhering to what the Holy Father teaches" and faith in the individual propositions contained in this address, then such a Catholic could take refuge in the decision of not wanting to know exactly what the Holy Father is saying, so as to avoid having to make an act of faith in what he is saying. This would prevent a Catholic from coming to the defence of either the person of the Holy Father or of his teaching. All that such an escape from reality would do would be to act as a kind of 'cushioning'. However, it is certainly not thomistic to avoid precision-of-knowledge to obviate having to make painful decisions about individual sentences read out by the Pope, allowing a blanket principle, the one of blindly adhering to what the Holy Father proposes, to be wrongly invoked when closer study would have revealed that in this case the Holy Father had something different in mind. Faith in abstract principles is alright, but this breaks down if they are being invoked (preferred) as mere expediency, so the need of having to deal with a painful but concrete reality can be avoided.

I am fiercely convinced that lies will never be grafted on my Vine either by the Holy Spirit or by the Holy Church or by the Pope. And so I could safely admit the Holy Father within the confines of my hedge, and sit with him, going through his Galileo speech line-by-line without any fear that any attempt would be made to graft onto my Vine what previously had been held up by the Holy Church as being not true. And so it simply became a very interesting and very precious mental exercise. And the Pope's message was:

> *This is the package I received from Cardinal Paul Poupard and from my Galileo commission. If anyone else had given it to you, you would have rejected it for its obvious bias and lies. But now that I have gone through the humiliation of reading it out to you, now I can tell you 'Watch out'! For, going by American and European voices, these are the things which Teilhardians find important, and these are the things by which innocent souls will be seduced to leave the Holy Catholic Church to find themselves*

Epilogue

transferred into the One-World 'Church of Darkness' of which my holy predecessor Pope St. Pius X spoke in 1910 …

And since *this* was the truth, I allowed him to graft *this message* onto my Vine…

Many examples can be quoted in John Paul II's pontificate of this highly successful indirect way in which this Pope deals with explosive issues. It is certainly not out of character (and I could almost add: for this accomplished actor) to be quite deadly with what looks at first glance to be no blow at all. We may be allowed here to point to the action the Holy See took when RENEW was at its height. Had the Holy Father met the deadly poison head-on there would have been endless bickering between the Papacy and the National Episcopal Conferences over which aspects of RENEW the Holy See took exception to. Instead, at the height of the oil slick, the Vatican issued a little study called *Report on Sects, Cults and* <u>*New Religions Movements*</u>, which quite clearly contained the message:

> *Catholics of the world. If you find operating in your neighbourhood a sect, cult or '<u>new religious movement</u>' with the objectionable characteristics as set out and discussed in this Report, be on your guard and have nothing further to do with their aims and objectives <u>nor with their recruiting methods</u>.*

It killed RENEW without even once mentioning the program by name …

We may take it then, that even in his *style* of teaching, our present Holy Father is guided by the Holy Spirit.

Finally, what must be our response?

Here, as good Thomists, we must first make a distinction. It is very necessary to distinguish here between *method* and *content.* Even if the Holy Father's *style* may at times be indirect, his *message* is anything but! We must not forget that actor Karol Wojtyla got his training in being *direct* by *indirect* means in identical hostile surroundings in which now Pope John Paul II finds himself. A training in method and style to pass on vital information, life-giving truths ...

If someone can be as direct as this Holy Father can be by indirect methods, let him or her go right ahead, as long as it does not adversely affect the message of truth. In the Pope's case we have that guarantee. Have we got that same guarantee in anybody else's case?

If Weakland quotes the Holy Father's Galileo address to make an unholy breach for a patent untruth, it will obviously lead to his own downfall. For the Holy <u>Father has been protected by the Holy Spirit against having *taught* what Weakland claims he taught</u>. In anyone else's case it would be very difficult to establish the same degree of protection by the Holy Spirit, if indirect methods of teaching were being applied. If the Holy Father holds up a lie, we have the guarantee, *which nobody else enjoys* that he is not teaching it, but is merely quoting it for some other very obvious reason which is bound to come to light after careful study, which will make it plain why a direct rebuttal would have done more harm than good.

This careful study we owe to him as his dear and faithful children.

Epilogue

It is therefore altogether impossible that a Pope in one part of an address (line *10*, in the non-Galileo part of it and in complete harmony with all previous papal teaching) can hold up that it is illogical to attribute to the *nature* of science the competence even to understand its own place in the nature of things, and then in another part of that same address (line *33*) can attribute to the *nature* of science the competence *"to limit the message of Faith"*. We must therefore conclude against Weakland, Poupard and their Teilhardian friends, that line *33* was ***not*** taught but only read out as 'black'.

With this I hope I have put the finishing touches to the very difficult question of the papal Galileo speech.

www.ingramcontent.com/pod-product-compliance
Lightning Source LLC
LaVergne TN
LVHW020936090426
835512LV00020B/3390